COLD ANTLER FARM

COLD ANTLER FARM

A MEMOIR OF GROWING FOOD AND CELEBRATING LIFE ON A SCRAPPY SIX-ACRE HOMESTEAD

Jenna Woginrich

ROOST BOOKS

BOSTON & LONDON 2014

Roost Books
An imprint of Shambhala Publications, Inc.
Horticultural Hall
300 Massachusetts Avenue
Boston, Massachusetts 02115
roostbooks.com

Illustrations on pages ii, 25, 115, 163 by Jeffrey Thompson

9 8 7 6 5 4 3 2 1

First Edition
Printed in the United States of America

⊗ This edition is printed on acid-free paper that meets the American National Standards Institute z39.48 Standard.
♻ This book is printed on 30% postconsumer recycled paper.
For more information please visit www.shambhala.com.

Distributed in the United States by Penguin Random House LLC and in Canada by Random House of Canada Ltd

LIBRARY OF CONGRESS CATALOGING-IN-PUBLICATION DATA

Woginrich, Jenna.
Cold Antler Farm: a memoir of growing food and celebrating life on a scrappy six-acre homestead / Jenna Woginrich.—First edition.
Pages   cm
Memoir of growing food and celebrating life on a scrappy six-acre homestead
ISBN 978-1-61180-103-3 (pbk.: alk. paper)
1. Farm life—New York (State)—Washington County. 2. Women farmers—New York (State)—Washington County. I. Title. II. Title: Memoir of growing food and celebrating life on a scrappy six-acre homestead.
S521.5.N7W63 2014
635.09747'49—dc23
2013038405

CONTENTS

The Kids Are All Right: A Prologue   I

The Leap Year: An Introduction   11

Spring: Uprisings

Before the Thaw   27

Cold Crops   37

Lambs on the Ground   41

A Rhapsody of Basil   48

Ostara   54

Delivery   57

Code of the Pig   61

Sheep School   68

The Life of Hay   71

Riot   78

Howlers   80

Rogue   83

Shearing Day   89

Blossoming   94

Tour of the Battenkill, Farm-Girl Style   97

Black Leather, Saddle Leather   106

Fire of Beltane   112

Summer: Growth

The Breakfast Club 117

Greens 124

Hardening Off 128

Weeding 132

Fences 135

Windfall 139

Blessed Humidity 141

A Compassionate Harvest 144

Homesteader's Vacation 152

Autumn: Harvest

Lughnasa 165

Dinner Date 172

Searching for Ships 175

Mabon 179

Sweater Song 186

Hunting 192

Hallowmas 196

November Fire 202

Sighs 205

Quiet Light 210

Luceo non Uro: An Epilogue 213

About the Author 216

And when the children are safe in bed, at one of the great holidays like the Fourth of July, New Year's, or Halloween, we can bring out some spirits and turn on the music, and the men and the women who are still among the living can get loose and really wild. So that's the final meaning of "wild"—the esoteric meaning, the deepest and most scary. Those who are ready for it will come to it. Please do not repeat this to the uninitiated.

—GARY SNYDER, *The Practice of the Wild*

THE KIDS ARE ALL RIGHT
[A PROLOGUE]

IT IS NOT EASY to balance a sixty-pound bale of hay while walking up an ice-coated hill. It's certainly not made any easier when a young border collie is circling your legs chasing a particularly nasty rooster. But here at Cold Antler Farm chaos is a standard practice. And this is where I find myself on an otherwise normal round of morning chores. Besides the dog at my feet there's a gnashing wind, making the entire mountainside seem loud and angry. I squint— the harsh winter morning sunlight reflecting off the frozen white ground makes it hard to see. I balance that hay, and avoid tripping over avian or canine, while wondering if the farm is causing me any premature blindness. I smile and turn my head down a bit, looking for proper footing as I crabwalk past a slippery spot. I would be annoyed at my dog if my brain had the capacity to focus on irritation and staying upright at the same time. It doesn't.

My horses are watching from behind their gate, leaning over it and making those rumble-throat noises that translate perfectly into English in any equestrian's head. We all know our mounts, and know that this is after-hours-comedy-club-level heckling. The reason I'm on this death march is to deliver their morning rations, and they clearly feel the service at this place is lacking. The geldings are pawing at the gate, excited about the salad bar in my arms. They are at the entrance of their winter quarters, a large fenced paddock with a shelter built into the side of a mountain. It's a scrappy setup, but it suits this farm and its purpose. The woven-wire fence and metal gate keep them in (most of the time), and covered with ice and snowflakes it looks downright lovely. With their bobbing heads and whinnies, sending clouds of smoky breath into the air, I don't even notice the broken T-posts. They are charming in their utility.

I'm in fairly good shape, but this day has me huffing and puffing. Right as I start to break the bale apart into smaller handfuls and feed the scoffers, I feel my phone vibrate in my pocket. Too tired to bother to answer it, I keep handing flakes of hay over the gate to the boys and call out to the sheep who are watching me from a high pasture. They know they eat at the same time as the fat ponies and they start their own chorus, asking and telling me at the same time that they are hungry too. The phone rings again, and I check the number. I don't recognize it so I decline the call and shove the phone back into my front jeans pocket. I grab the two empty buckets by the horses' water trough and start the obstacle course back down the hill. The phone rings again. Now, a bit peeved, I pick it up and answer, probably not as politely as I should have.

"Hello?"

"Hey, Jenna?" I instantly recognize the voice. It's my good friend Yahsh, who runs the dairy-goat operation three miles down the road at Common Sense Farm. Yahsheva might be the kindest, gentlest, and most patient person I have ever met. She's in her late twenties, a mother of four, and wife to the hardest-working farmer in Washington County. Despite all that, she glows in a way few women can past the first months of motherhood. She's a goddess, quite frankly, and she doesn't deserve my terse tone. Especially considering the news she is about to deliver.

"I'm in the barn with Iris and she's in labor. I think she'll be having twins, so if you can make it here in the next ten minutes you can see what a birth is like and how we get the new kids ready for the farm."

"I'm on my way!" In those four words I go from agitated to apologetic and then purely elated. Being the recent owner of two fat and happy does about to give birth in a few weeks, this firsthand experience with caprine midwifery is a gift beyond measure. Yahsheva knew I wanted to be present for a healthy goat birth and she called from the closest phone she could find (which is why I didn't recognize the number). I finish my chores in record time, feeding the flock of sheep, the goats, horses, chickens, and rabbits, and wrangle the reluctant rooster back into the chicken yard. I grab a sweater and a camera and hop into my dented Dodge pickup. I turn the key, the engine coughs into a proper start, and I practically peel out of the icy driveway, heading due south.

Common Sense Farm is different from most working farms in the area. It maintains a solid CSA (community-supported agriculture) program that makes vegetables available to its members, and its residents belong to a religious commune that emphasizes voluntary simplicity. When I moved here I immediately showed up at their door and introduced myself, and I was blessed by a seemingly endless supply of kind neighbors and new friends. I became close friends with Yahsh and her husband, Othniel. Over the years we have helped each other on our farms and traded equipment; shared rides to feed stores, and many meals. I feel a strong sense of common purpose with them. I see Yahsh's family more than I see my own, and I know their farm and estate better than most people know their town's public parks. Their commitment to sustainable agriculture, organic dairy work, and the larger community inspires and supports my own.

I drive faster than I should. It has been a harsh winter—there wasn't much snow but plenty of bone-deep cold and ice. I am on the little highway—Route 22—that connects our farms. Well, it's one of the ways we connect. I live just two or three miles north of Common Sense, and we are separated by a few small foothills of the Taconic Mountains. So besides the conventional automobile roads there are unpaved roads where I can take a horse cart or saddle up and ride. Just a few paces from my driveway is an old logging road that connects my mountain to its smaller neighbor. That smaller hill connects to another large farm estate that neighbors the commune. I realize that it's the twenty-first century, but I am very happy to know that there is a network of paths my horse and I can walk along that lead me to other like-minded, fearless agrarians.

To have friends who understand that, live that, and are happy with that is so encouraging to a greenhorn farmer like me, just starting out. Like minds are a rarity, and we appreciate fellowship where we can find it. The Common Sense folk are similar to the Amish but accept many more modern conveniences. I like knowing that a band of what I nicknamed techno-Amish with an endless pool of farm resources is just a few miles away. You lose your preferences for the conventional when you need a syringe of selenium at midnight for a ewe shaking with toxemia.

When I reach the large barn at the back of the commune's vast estate of about two hundred acres I am greeted at the barn door by one of Yahsheva's children, a tow-haired five-year-old with a personality as big as the barn. Her name is Emmith, but I call her the border collie. She is as tenacious and energetic as any sheepdog could ever dream of being, and already is just as smart in her own way. The group doesn't allow pets—you are dealing with seventy people with dozens of family units in one place; if every family had a dog it would be a kennel as much as a commune. Not to mention issues like allergies, dog bites, and so forth. So there are no canines on this farm for several reasons, but even so, I don't think anyone took that nickname as an insult. They know how I feel about good dogs. Hell, some of the best people I have known were sheepdogs.

Emmith—with long flaxen pigtails and wearing a pair of mini-coveralls—grabs my nervous hand and leads me back to the birthing pens. I trust this little girl. I watched her grow up, and if anyone can lead me to her mama and the scene of the day, it is this pup.

A crowd is gathered, at least fifteen people. It's mostly women and their children, done with their day's lessons in the schoolhouse and ready for their real education here in the kidding sheds. New goats are a big deal on a farm that specializes in dairy. It is like a holiday celebration, the kind of thing mothers want their children to witness. A teenage girl is already toweling off the first of Iris's kids, whose birth I just missed, but there is another one on the way. The girl is dressed for barn chores, but without deviating from the standards of modesty the group suggests for their people. She has on loose-fitting baggy pants that could easily have once been a long skirt cut up the middle and sewn into two halves. The skirt-pants are tucked into rubber boots, and a canvas jacket covers her wool sweater. Her beautiful blond hair is tied back in a ponytail. In fact, she looks like a young farm woman from another time, the Civil War era perhaps. A time when conventions of fashion met the reality of the work and women's wardrobes adapted with the efforts.

Yahsheva walks up behind me in her loose-fitting wool sweater and those commune-issue baggy pants. She gives me a hug and pulls back, all red-cheeked and smiling. She points to Iris, the black-and-white Alpine doe in the pen (Alpine is a domestic goat breed that originated in the French Alps). The new mother is standing up, all four legs tense. Out of her rear end a balloon of fluid, about the size of a large orange, is emerging. It's a birthing sac, full of water. Several young husbandry apprentices are with us, and Yahsheva reminds them to pay attention. She reaches over to the swollen fluid sac and bursts it with her bare hands. The fluid that pours out is clear, but there's something else in the sac, something possibly ominous. Streams of yolky yellow ooze out and cover Yahsheva's hands. Neither she nor I so much as flinch. The crowd around us isn't bothered either. All of us here are used to this messy business. I have been at the side of more than a dozen sheep births, and she has been midwife to more goats than she could count. But her expression does change. She looks worried at the sight of the yellow fluid dripping into the hay. The contents of the doe's burst sac look like a giant cracked egg. It is a disturbing image, and one she has seen before. She tells me how last year Iris had a discolored sac, and only one stillborn babe emerged, already dead inside its mother. It was a truly disappointing day, and emotional for the children. She sighs and warns me that this could be unpleasant. I nod, ready for the worst. Not all of farming in the spring is seedlings and happy baby animals. As the saying goes, where there is livestock, there is dead stock. There was a good chance this second kid could already be dead inside the womb. Everyone is prepared for it, and Yahsh decides the best thing to do is get it over with. The bleating mother, the crowd, and her newest student (me) just want to get past the tension. We can see two little hooves poking out of the birth canal, and May, the same apprentice who was seeing to the first kid, bends over to get a closer look.

"Two hooves and a mouth! The tongue is out!" May exclaims happily. I'm not sure whether her joy should be tempered, as newborns with tongues hanging out is not a good sign, certainly not in cattle or sheep. Yahsheva steps in and carefully slips her hands inside the doe to check that the kid's feet and legs are in the proper position for easiest delivery. Certain they are, she tells May to pull the babe out. With a gulp and confident grit, May does as she is told and pulls the little black-and-white kid out of her mother's body in one adept motion. If you had to tie a rope around a toaster-sized serving of canned dog food and pull it out of the can, that's how long it takes and exactly how it sounds. Within seconds of this new life's hitting the hay, Yahsheva and May dive in to remove any mucus and sac remnants from its mouth and nose. The little babe gasps for a breath and starts to move, and smiles fill the barn with wild light. Yahsheva lifts a soggy leg and tail and can hardly believe their luck; not only is the little one alive but she's a girl! Everyone is beaming at the news, thrilled to have a new little doe to join the throng of newborns in a hay-bale-lined baby pen adjacent to the birthing pen.

I look around at all the glowing faces, and everyone is encouraged by the new additions. Toddlers and grandmothers alike are whooping and smiling, holding the fluffy pair of kids born earlier that day in their arms. There's a celebratory feeling all around. This proud mother who had a single stillborn the season before has just brought a pair of beautiful twins into the world. Talk about an increase! As this newest member of the farm is toweled dry and cleaned up, she is whisked away from her mom. In the milk room she is gently blow-dried by adoring future farmers. This may seem cruel—the mother and kid having zero time to get to know each other—to those who are new to working dairies, but the reality is that most of the milk from these goats is for the people who farm them, not for the baby goats. Lactating Alpine does produce far more than the little ones need—they were bred for just this purpose. The humans who feed and house them get most of the milk, and the little ones get bottles instead. They are fed plenty that way and get to learn the Tribes folk as their imprinted authority figures and family. It makes for a more manageable herd and a more productive dairy. I am thinking about this as the soft hum of the hair dryer behind a wooden door lulls me into thoughts about my own goat kids just around the corner. Where will I keep them? Do they need their own pen in the barn? Should I get a furniture box from the hardware store, line it with hay, and keep it in the same room as the woodstove? I'm starting to realize my lack of planning and I want to ask Yahsheva all about it, but her attention is back on Iris. The mother of twins bleats while another birthing sac pops out of her rump. *Triplets?* Clean towels are gathered, hands are washed, and we all hunker down for the big show.

The third kid arrives moments later, followed by thick, red strings of placenta. It's another girl, and Yahsheva is grinning from ear to ear. Out of six kids, four new does were added to the farm this kidding season. Those are some beautiful odds in a small dairy.

Iris is finally done, and as the last of her babes is cleaned up and taken to the hay-bale kid pen she is given a bucket of electrolyte-spiked water to replenish herself. While she gently drinks you can almost sense a release of tension, the calm returned to her. She is then given a little grain from a pan that rests on a hay bale at her head height. While Iris inhales the molasses-coated calories Yahsheva swoops in with a milk pail and expertly milks out half of her udder. I ask her why only half, and she explains that a full milking would be too much on the mother, as no newborn in nature would milk her mother empty on the first day of life. She takes the milk pail, about a quarter full of the eggnog-like colostrum, which will be poured into bottles for the triplets. This is the proverbial mother's milk babies so depend on. That colostrum is rich stuff, full of all the antibodies and nutrition a babe needs to ward off hunger and disease.

It's been less than an hour, the barn is three goats richer, and—like everyone else who was around the little kids and the birthing pen—I am stained with placenta and goat feces. Not one of us cares and we share some happy hugs. None of us has gotten over the success of the morning, except Iris, who quietly sips her water. I quietly note that it is a few days after Imbolc, the blessed midpoint between the winter solstice and the spring equinox, and a festival celebrating fertility.

I am trying to think of what success meant a lifetime ago, when at this time of the day and week I would be buried in a web design program in an office to which I commuted every day from my farm. I designed websites for nearly eight years after graduating with a BFA in graphic design, but in 2012 I left the job for this extremely local life as a farmer on a mountainside. I didn't really have the resources to quit my job—money has always been tight—but I did so because I was extremely unhappy. I watched too many people decades older than I schlep through their day jobs with dead eyes and broken hearts. I refused to be one of those people; besides being fearful, I knew I did not belong there. I wasn't good at my job, not really. I did the work that was required to get a paycheck and I spent nearly every day in the office performing out of fear of losing my job (and farm), and not any sort of joy of the work.

Eventually I could no longer wake up and leave my farm for that place, so I resigned my job. I cashed in my meager $9,000 401(k) plan and decided to make a living teaching, writing, and running a website. I have been *just* getting by ever since. Some saw it as an act of bravery, others, as an act of pure stupidity. Either way it was my choice, and I'm glad I made it. Even though I only commuted twenty minutes to work in the office, on days like this, where wet new life is just minutes away, those twenty minutes feel like a cross-continental flight. Today my personal, professional, financial, and gastronomical life exists mostly within a ten-square-mile area. That old commute feels like a legend from another time. And as I look around this sun-kissed barn—covered in blood and poo—someone hands me the just-born triplet. She is clean and her coat is dry and her eyes could melt the heart of the most miserable person in the world. I smell her scruffy head, feel her shiver under the blankets, and whisper to her until my body heat stops her fuss. She closes her eyes and rests her head on my forearm and patiently watches as her siblings sip from the bottles. I am handed a bottle and I offer her that first meal, and she greedily accepts. The barn is quiet now—even the human toddlers seem in awe of the babes sucking from bottles and the now-restored calm in the old barn.

How could I ever not have chosen this?

THE LEAP YEAR

[AN INTRODUCTION]

WHEN I GRADUATED FROM COLLEGE in 2005 I had wanderlust in spades in addition to my BFA. This is a powerful combination in a robust economy. Jobs for designers abounded, and I was willing to travel to work. I wasn't alone. My peers were headed for jobs in nearby cities such as New York, Philadelphia, Boston, and DC—to live in trendy neighborhoods such as Williamsburg and Cambridge—Massachusetts, that is. I thought I'd end up there, too, but when fate came a-calling it was the South that hollered my name. I was offered a full-time gig in Knoxville, Tennessee, working for a television network, and it was too good a chance to pass up. So I said good-bye to dreams of Madison Avenue and headed to the Smoky Mountains.

To summarize nearly two years of adventure and increased self-awareness: I realized that my move to the mountains was more than just a new job. It was a paradigm shift that changed the course of my life. I fell in love with Appalachia. I loved the people, the music, the history; and I fell head over heels in love with those rolling mountains and babbling brooks. I spent more and more time among them. Every weekend I would scramble into my Subaru with my dogs and head west to the trails I now felt more at home on than my city block. No part of me wished for Madison Avenue anymore. It was as gone to me as Pharaoh's Egypt.

One of my favorite adventures was a drive around Cades Cove. Cades Cove is a verdant valley in the Smoky Mountains, surrounded by an eleven-mile loop that was a once-thriving nineteenth-century mountain community. It has a historic village complete with cabins, farmsteads, a grain mill, and a white church with simple wooden pews. By the time the likes of me walked around that sacred land it had long since changed. The place that had once echoed with the sounds of mule wagons and music coming from handmade fiddles was a tourist attraction. You walked around to try to understand what once was—like looking at the bones of a prehistoric beast in a museum. The present Cades Cove was a display piece now, but the honesty was there. I hummed with wonder.

That time spent in Cades Cove and on the trails gave me a quiet thrill. I started learning about the original settlers who lived there before the federal government took it over as part of the Great Smoky Mountain National Park. Their story of community and survival was nothing short of heroic. These people could make clothing out of sheep's wool! Could make dinner out of a garden and a whitetail deer! These people knew how to live, *and live well*, on their own. I remember thinking: if it weren't for my car and a grocery store I would be helpless. My wonder turned into unease.

So in the spirit of independence, I decided I too wanted to learn those skills. I wanted to eat in more often, to provide for myself. To me that meant growing at least a few items I could consume, for the novelty if nothing else. I didn't think I had the chops to sew myself a pair of pants yet, but I could at least learn how to sew on a button. And I was pretty sure I could decipher a bread recipe. I tried these things and they excited me. Those first loaves of bread from my own oven created a doorway into another world. I added a pot of snap peas to the windowsill, and sewed a pair of mittens from a tracing of my own hands. I didn't have to live the way I had assumed I did. I could reclaim some of what all the people I knew had lost.

That novelty turned into a mission, and I started devouring books about gardening and animal care. It was the animals that I was most drawn to. In my view anyone could grow a potted plant, but it took a certain level of success to have the space, time, and confidence to lead a goat on a halter down a road. I learned livestock breeds and how to tell a beef cow from a milk cow. What seems like children's rural education was the level I was working at. On the way home from the office I would stop at a bookstore and swig overpriced frozen coffee drinks and flip through *Hobby Farm* magazine. I sipped my corn syrup and stared at the people in the pictures, trying to get clues from their faces about how they managed to find their own Cades Cove and how I could get there too.

It was my second summer in Tennessee that turned this fascination with homesteading—living off the land, raising animals, exploring self-sufficiency on a small scale—into action. It was in the holy Cades Cove itself that an event occurred that changed my hardwiring. One muggy summer day my roommate, Heather, and I were hiking on the Abrams Falls Trail to visit the waterfall at the end of the path. Abrams Falls was not a waterfall you could drive up to and snap pictures of. It was a two-and-a-half-mile jaunt to get there, but it was worth the effort. At the end of that journey there was a beautiful twenty-foot waterfall emptying into a trout stream of cool mountain water. After a few miles carrying daypacks loaded with lunch, towels, and our regular day-hiking gear we could not wait to jump in for a lazy soak. We weren't alone either. All around the wide swimming hole were college students, families, and a few couples who took to their Saturday afternoons just like we did. I felt a happy camaraderie with the folks, all of them strangers but

akin in our idea that a five-mile trek was a fun way to spend a sweltering summer afternoon in Tennessee. The swimming hole was a
calm pool far away from the rocky edge of the falls. I took in the
moment, soaking up the laughter, sunlight, and people.

After a long walk in 100 percent humidity you can get drunk
just looking at clear mountain pools. I was panting at this point,
shaking with eagerness to strip off my sticky T-shirt and shorts
and dive in. The bathing suit under my hiking clothes was already
soaked with sweat, and when I eased into the cold water it felt like
a blessing. As Heather and I swam, we chatted and laughed. I did
some long breaststrokes out to the deeper area, a safe distance from
the rocky falls whose undertow could trap you, suck you under,
and not let you go. Heather was trying to catch with her hands a
brook trout that swam past her legs (no luck). We were both revel-
ing in the joy of it all when we were shocked out of our meditations
by a giant *ssssplash*. The source was a guy in his twenties who had
just jumped from a rocky ledge to the side of the waterfall into
the deep pool below as if he were jumping off a springboard at the
community pool. He popped out of the water grinning after his
brave leap. We watched several of his friends hike up to the top
of the falls, stand on the ledge, and jump off like it was the most
natural thing in the world to do. I was in awe, watching these ath-
letes from the safe shallow end. I was not what you would consider
obese, but not small. I had about thirty pounds of extra weight on
me and it was woefully distributed evenly around my entire body. I
felt like a little oak barrel watching Greek sculptures, utility versus
art. I wanted to feel like those young men. I wanted to know what
it was like to actually live life instead of watching others live it.

A wonderfully horrible idea came over me.

I wanted to jump, too.

What the hell, it didn't seem that high. And they all seemed safe and sound when they got out of the water. I was full of piss and vinegar so I looked at Heather and said, "Let's jump," and she shook her head no. (Heather is smarter than I am.)

Fine. I shrugged and decided I would go alone. I climbed up the falls from around the back and followed a very brushy but well-worn path. Heather started to follow me. There weren't any signs telling us to turn back. I watched as another guy made the jump just a few feet in front of me. I felt the air leave my body as anxiety ate me up. I had managed to forget my fear of heights in all that excitement, and now, from the top of the falls, those twenty feet seemed *a lot* higher. Too late now. I made my way to the very edge, barely hearing the screamed advice from the man-boys watching from below. They seemed to be pushing their arms out, telling me through universal flailing language to make sure I cleared enough space between me and the fall in my leap forward. I looked right below me and there were mean-looking jagged rocks. A hiker screamed up to me through cupped hands like a blow horn, "You need to jump out at least five feet or you're plasma!" I nodded—but when you are a shaking, five-foot-three chubby gal and scared of heights, flinging yourself five feet away from a rocky ledge seems damned near impossible. Something went off in my brain and I just went for it. I pushed out best I could. I jumped.

I didn't make the five feet.

I can still remember, clear as a still frame from a movie, seeing those rocks coming right at me. I remember closing my eyes in fear, feeling the useless regret. Then the slap of my body hitting water, and I was under. I remember the equal parts of pain, relief, and joy when I realized I wasn't dead. Still a few feet under water—not close enough to the undertow to be consumed by its wrath, but not far from it either. I opened my eyes in the dark brown-green to see myself still whole. Spending time under the cold water felt like a choice. If I surfaced I might see the blood or realize my spine was split in two. So I just hovered there, floating under the rushing sound of the falls to my left without time or space or emotion influencing me. I didn't think about air. If I had to breathe, it wasn't enough to snap me into instinct. I remained incredulously suspended, amazed I still could think. I said my name underwater, making sure I knew so much.

Only when I finally swam to the surface did I realize I was not dead. I looked around grinning, and I now understood what that smiling guy knew. I was the only person with a bit of serenity. A few of the guys were standing on the edge of the rocks, all of them pale. One stuttered, "We thought for certain we'd have to go in after you . . . you missed the rocks by this much," and he showed me an inch or two of distance between his fingers. I started shaking then, deep under my skin. I pulled myself out of the water to the safety of the giant flat slabs of river rock where my daypack awaited me. I curled up in my towel like a wounded bat wrapping herself up in her wings and sat down. From there I watched Heather leap off the ledge and clear the danger zone with aplomb. I couldn't help but whoop for her, even though my ribs rattled my suet. I felt as horrified as I felt validated. My humble hobbit body radiated pride.

No one else seemed to want to jump much more that day. Heather and I packed our bags and hiked back to the parking lot. I shook at a low frequency the whole walk back to the car. I had never been so scared and happy before. Never felt more awake. Never felt a truer gratitude than I did on that familiar trail in the Smoky Mountains. It could have all ended on the rocks, but I had been spared. Two big crows cried out from a tall pine, and I watched them fly down the mountain with us. If they had seen our foolish leaps, they refrained from judgment.

That fine day I got a lesson from those mountains, from that jump. It may be the reason I am a writer and a farmer today. It made me understand that although most leaps of faith are dangerous, what could be waiting in the cold water might be more than a hiccup in time—it could be a baptism.

The jump was certainly stupid, but I was proud of surviving it. It felt like I'd cheated death, achieved the great escape. My smug pride was short-lived. A few days later I read the news that a few college students on holiday had gone for a swim in that same pool and one had drowned. He didn't jump. He just swam too close to the falls, and the force of the falling water's undertow pulled him down to a depth only a diving team could rescue his remains from. I had jumped from a deadly height and survived without a scratch. He doggie-paddled too close while taking a picture with a waterproof camera and was dead moments later. This is an unfair pair of stories, as arbitrary as leaves falling from a tree. I understood, theoretically, that my life can be taken at any time, but never before had this truth rung so loud in my head. I had my story and newspaper clippings about the other one to prove it, tangible evidence that only luck and a pair of crows filled the air between two lives, one still present and the other gone.

Five months later I was out of Knoxville and on a rented farm in Sandpoint, Idaho. I had been offered a job working for a clothing retailer in the Northern Rockies. They paid for my move, and once again I drove a packed station wagon to my new home. In hindsight, it was an erratic and crazy change in my life, but I am certain that surviving that jump was the only reason I could pack up from the city and move cross-country alone. If I made it that time, I could make it again. So I left the city and everything I had built, career- and relationship-wise, to follow an idea I understood only from the vantage point of a bookstore café's magazine racks. For me, it was 2,800 miles and three time zones away to turn glossy pages from a farming magazine into reality.

Idaho was pure country, and it demanded that I allow change to enter my life. I had a barn now, places to garden, and folks haying outside my backdoor. Just being there gave me all the permission I needed to set up a chicken coop and beehive and buy fiber rabbits. Sunflowers grew along the red fence posts, and I entered my pet chickens in the county fair. I was no longer the graphic designer going to her local AIGA meetings. I was a "farmer," or the beginnings of one. The person on the earlier side of that leap disappeared into obscurity, and an updated version who could play the mountain dulcimer and knit a hat showed up. I liked her more. I let her stay.

Idaho was a dramatic place, and it allowed me to make dramatic changes. I started writing every day about my country-living adventures and kept up a blog for friends and family on the East Coast. A friend of mine in Maryland suggested that I write a book about my lifestyle. So I asked Storey Publishing if I could pen a book for them about learning self-reliance as a single woman and farmer in her twenties. Ten years ago this was a new idea, as there were no books on backyard farming, and "urban homesteading" was a phrase only spoken on the fringes of society. I jumped again: I signed my first book contract on my twenty-fifth birthday.

If it had not been for that hallowed day in Tennessee, Jumping Day, I'm certain I'd be another person entirely. I would have put aside the dream and focused on my career, the dating scene, the newest Indian restaurant near the college center. The jump took away the fear of my own early death to stop putting off life. It took a reaction to panic—and I know this is not special nor particularly healthy—to shake the logical out of me and send me out West on a lark. If I hadn't jumped, I'd still be just another hipster with a wall of musical instruments on display I never played and a useless stack of farming books by my bed, the pile for "someday."

I adored Idaho, but I ended up missing Tennessee very much. I think in a lot of ways East Tennessee was my first true love. A place that I got lost in, that changed me, that I had left at the peak of our intensity: the perfect recipe for sappy regret. The stoic gray mountains, moose tracks, and deep lakes of Idaho could not compete with the romance of those Southern mountains. It wasn't that Idaho wasn't beautiful—I was just spoiled on one place and missing it so harshly that it could crack a rib. I watched movies that took place in the South to satiate my desires, and documentaries about Appalachia. I was a junky looking for another hit.

And all of those movies, the writing, the nights staring at pictures of hiking trails a continent away, they stirred up a new longing in me. I wanted to reconnect to the people, the mountains, and the back roads to Asheville, North Carolina, that I still knew by heart. What connected all of this was the music: mountain music. I wanted those mountain airs and ballads in my Idaho farmhouse. That became the remedy for my longing, what could turn Idaho from an adventure into a mission. If I learned that music I could share it with new friends, have the joy of both places at once. So, knowing nothing about music, I bought a cheap violin off eBay and some instructional books and CDs and taught myself to fiddle. It was something I had always wanted to try, but I was scared to take on such a hobby. After Abram's Falls it seemed as simple a thing to take on as walking upstairs; you just go one step at a time. Not that the music or learning the techniques came easy, but my determination and constant sawing lit the way.

The Smoky Mountains remain home to me in many ways, no matter where I hang my hat. It is the place that made me face fear, accept death, and choose to keep living until it came. Those mountains made it clear that waiting to live the life you want is a ridiculous and dangerous luxury, and waiting to make a change is just taunting fate. You could be 9.7 years into your ten-year plan and get thrown under a bus. Sometimes it takes real rocks and water and a stranger's death to get over the clutter in your soul to make a change. It did for me. I realized the most valuable gift I had, the only gift I had, was the time that remained. So I jumped.

I moved to New York state after two years in Vermont. I loved the state, the little cabin, and the animals I raised there—but it wasn't my land. It belonged to a woman in Connecticut who used it as a vacation home, and my lease ended when she returned for renovations.

With a little luck, a special USDA program, and a wonderful mortgage broker, in the spring of 2010 I moved into the home I christened Cold Antler Farm. The 1860s farmhouse had my name on the deed just three months after the day I first saw it. Sometimes things simply work out.

And sometimes there's a reason: the house wasn't very popular on the local market. Families and professionals weren't interested in the old, wavy floors and the single bathroom on the first floor. I can't blame them for not investing in exactly what they wanted, or envisioned. However, I was not buying with such concerns on the table. I didn't need something for a magazine shoot or a remodeled kitchen with steel appliances. I needed land, water, grass, and possibility. The little house was a blessing, a perfect fit. I didn't care about my furniture sitting level, but I did care about the pasture, barns, forest, pond, stream, and outbuildings crying for a caretaker. It became home the moment I pulled into the driveway.

When I changed how I lived my life I also changed my calendar. It was possibly the most intense shift of thought that I undertook, and nobody prepared me for it. I grew up going to school, and then work, on weekdays and fitting all recreational, appointment, and social time into afterhours or weekends. I consider this the normal calendar for most people out there paying taxes and licensing their dogs. It's what a respectable citizenry does. But when I left the office gig the days of the week started to melt together—less about numbers and more about the work that filled those arbitrary shapes and names of the days of the week. Let me try to explain:

We live by the calendar. Regardless of who you are, how much money you make, or where you live, your measurement of time is the year. Whether that means you're cranking up for a publishing deadline or corn shucking, your timeline is based on twelve months of constantly rotating thirty-day cycles you will never escape, even in death. When you die people will mark it by the day and month, as they did your birth. We are an animal that only understands time through numbers. The notion that dates aren't related to time is nearly inconceivable, as undisputable as rain.

I was born on July 10, 1982.

There's a thunderstorm in the distance.

If those two sentences both sound like factual statements, then you understand my point perfectly. However, that first sentence we made up. Humans decided how to measure cycles on earth, and this use of seconds, minutes, hours, and days was the system that proved itself over time. It is handy for record keeping and for marking holy days and rites of passage. It isn't real like a thunderstorm, though. Go back far enough, before we spoke to each other, when the idea of agriculture was as far away in our primitive minds as Disneyland, and you understand. Earth invented thunderstorms. We invented time.

Thunder and lightning are beyond collective systems of time-keeping. Walk outside in a storm and feel the world shake, light up, and your body get wet as if it had been tossed overboard. Feel four ancient elements—earth, air, fire, and water—slam into your life at the same time. This real time comes and goes but it isn't marked by numbers. It is marked by experiences.

It took living on a farm for a whole revolution of the wheel of the seasons for me to truly know that in the marrow of my bones. I only tend a few acres of land, but it is land where for me time has returned to more primal systems of measurement, and that may very well be the land's greatest gift to me. My new reality is not anchored by hours and days but instead by systems of birth, love, sex, death, dirt, blood, and rebirth. Become a farmer, any type of farmer, and time changes on you. You don't get more of it, but it does become denser. You start to feel changes in the earth rather than tracking them on a calendar.

If you live an agrarian life in a climate like mine, you can throw away the twelve-month system and instead tune in to four quarter-turns of the wheel: winter, spring, summer, and fall. Four seasons become your new religion, regardless of your beliefs. You tend to them, act in regard to them, and owe your life to them—because, while you may be new to agriculture, the human race certainly isn't. Join the ranks of the ages, hold up your hoes, and discover that your birthday, or "Saturday," or "New Year's Eve" loses meaning. Holidays change on you the way a deer's coat morphs from red to dun in the winter. There isn't youth, middle age, and retirement—there is just alive. It's the kind of reality you only start to understand when waterfalls and thunderstorms soak you thoroughly, and daylight is no longer doled out in hours and minutes.

This book traces the shape of a circle—a wheel, really. The old agrarian calendar used to be called the Wheel of the Year. Today modern revivalists and reconstructionists of the old holidays still call it that, even if the nomenclature feels fuzzy. Regardless, a wheel is a fitting symbol, since it is such a utilitarian representation of continuation.

To help navigate my stories, picture the Wheel as a wagon wheel with eight spokes. The four main spokes are the seasons: spring, summer, autumn, and winter. They are anchored by the solstices and equinoxes. Bisecting the quadrants of the wheel are the cross-quarter days—the important pre-Christian agricultural days of ancient Europe: Imbolc, Beltane, Lughnasa, and Samhain. The modern incarnations of these days have names such as May Day and Halloween. By this timeless Wheel, this farm's story can be best told.

SPRING

UPRISINGS

BEFORE THE THAW

I PAY ATTENTION TO THE WEATHER more than anyone I know. My health, my work, even my social life—all these things are contingent on the forecast. It is beyond ritual or logic, this weather checking. I own meteorological gadgets such as few normal citizens own. I have rain gauges and barometers and remote-controlled indoor/outdoor setups that compare and contrast conditions from my barnyard to my living room. When I wake up in the morning the first thing I do—before I take off the covers or consider a trip to the bathroom—is check the weather. In the dark cold of a winter morning a brave arm reaches out from under my warm blankets and clamors for my smartphone. That phone is always close to me, and not because of texts or Instagrams but because when you live like I do, those weather apps become your babysitter, best friend, worst enemy, and fortuneteller. Within seconds of clicking on my news I find out everything that is happening to me that day, the next day, and probably into the rest of the week. I would call myself obsessed, but that isn't accurate. Obsessed implies some sort of control. I am possessed.

I, Jenna Woginrich, am weather's bitch.

I didn't always live like this. I can remember—just a few years ago—the days of waking after hitting SNOOZE at least four times and staggering like a zombie to the coffeemaker. I'd shower and dress myself, and only when I realized I was about to walk outside to my car did I even consider the atmosphere. And even then it was only for reasons of fashion. Was it a light jacket or heavy cardigan day? Should I avoid the suede boots in case of puddles in the office parking lot? That brief acknowledgment was all the import I assigned to weather. Meteorology seemed archaic to me back then, lost lore, something for someone else to worry about.

In the farmhouse, under a quilt and fending off the forty-five-degree (indoor) temperature, I am seeing a preview of just how cold it is outside. This is how I start nearly every winter day here in the upper Hudson Valley. If it says it is minus seven degrees outside and I managed to stay up late enough feeding the woodstove so the house remained comfortably in the sweater zone of forty-five to sixty-five, I am ecstatic. Even before I brush my teeth I am a superhero. It is the morale booster I need to take on the day. My border collie, Gibson, is curled up so tight against my chest that he seems no larger than a throw pillow. When I lean over and whisper into his ear "Lessgo to work," his brown eyes shoot open, and out of the dense black round of fur explodes nearly sixty pounds of agile farm boy. He is off the bed and panting, his four white paws planted on the cold floor, and I swear that he is smiling at me. I see the swirls of mist rise up from his puffed-up chest. Gibson's entire life could be a trailer for the most inspiring agricultural movie of all time. The way he stands, solid as a comic-book hero, watching me and waiting for the holy moment I raise my sleepy body up and get dressed. He sees me reach for my glasses, for my oversized wool sweater, and he starts running for the door. He knows a world is out there, impatient to greet it. All farming border collies are loaded guns. Their ammunition source is endless.

When it is minus four degrees on a late-winter morning, my

work is cut out for me. By the time I am dressed and get my first taste of the weather, I am already making lists in my head. While the dog scatters about, releasing his morning deposits and checking the "message tire" on the pickup where every farm dog has left his mark, I am thinking about everything that has to happen before I can sit down with a mug of coffee. I live by the weather, and my needs are primal. After canine bladder emptying I must go through the motions of satisfying the needs of stove, farm, and self. This means first getting the chilly interior warmed up and a fire roaring so that when the outside work of feeding animals and checking fences is complete, I can come inside to a small gift of comfort and a place to hang my snow-wet scarf and gloves.

I grab the woodstove's ash pail and call the dog. We go inside together and I set down the pail by the stove and head into the kitchen to scoop a bowl of dog ration coated in a gravy of hot water. Gibson dives into the morning chow while I clean out a few days' buildup from the farmhouse's two woodstoves. When the stoves are ready for their new fires I'll get kindling and paper together and start the flames of the day. I gave up on heating with oil years ago. It was too expensive and too ironic—I mean, swilling fossil fuels to heat a sustainability-oriented homestead? I have forest, working horses, and harnesses. I use wood for fuel, and it feels correct. It just requires a little more sweat equity upfront; heating the house requires a few more wiles. . . . Now warmth isn't so much a turn of the dial as a game of chess.

On this particular snowy morning my mind is on spring, even in the throes of wood smoke and icy gates. I have seeds to order and chicks to reserve from a northeast hatchery, and the birth of kids and lambs is just weeks away. Soon I will be back to twice-daily milking of my big Alpine doe, Bonita. And as if that weren't enough of a challenge, I had just taken on a second doe, a yearling Oberhasli (another Swiss breed) named Francis. (Yes, Francis, not Frances. That's what the 4-H kid wrote on the pedigree form and I kept it.) She would need to go through her first birth and then be trained to stand for milking. I'm not sure either of us was prepared for that, but we'd find out soon enough. Twice the does means twice the milk, which means twice the soap and cheese, enough for gifts and bartering outside the farm—goat's milk is my currency.

While going about the morning chores, Merlin and Jasper whinny at me, as always, excited for their breakfast, though my mind is still on kidding and milk pails. My horses are tough stock, mountain ponies. Merlin is a Fell. He was brought over from England when he was four years old, and now at a mature sixteen, he is my riding and carting horse. At fourteen hands and nearly a thousand pounds, he makes people think twice about what the word *pony* means. The mustangs of the Wild West are the same height but a fraction of the weight. He is stocky and jet-black and has a long mane and a forelock that reaches the tip of his nose. His saucerlike feet have heavy feathering. He looks prehistoric, a mastodon, beside Jasper's smaller frame. Jasper is a Pony of the Americas, POA, a breed used for (mostly) children's Western pleasure riding. Jasper was my first horse, the one I cut my teeth on. He is trained to be ridden, and to pull and work if led by a halter, and he earns his keep. Just last year he pulled several stone boats full of locust rounds across the farm and to the woodpile. Important work while we were making room for the new paddock and pole barn. At only four hundred pounds and 11.2 hands Jasper is a shrimp next to the Fell, but he doesn't know it. Jasper is a fox, quick and agile and too clever for his own good. I secretly think he is the law in Horse Town.

My farm chores are wrapped up with those two horses diving into their flakes of green second-cut hay from last August. I take a minute to lean on their fence and tousle manes and tell them good morning. They are far more interested in the hay than in my petting but tolerate the gesture with patience. I turn around from the little hilltop gate and lean back into it, the two horses behind me eating and poking their warm wet noses at my hair and back as I take in the view. My empire is small. It still floors me that a whole life, a thousand adventures, and so much self-worth and hard work can fit into just a few acres, tools, saddles, and dirt.

Here I share my life with over fifty animals of various livestock breeds and manage six and a half humble acres—this compilation is my farm. With both critters and land in my keep I can no longer hit the SNOOZE button. It is the indisputable realities of temperature and precipitation that decide how I start and spend my entire day.

My whole schedule is waterproof—there is always plenty to do, regardless of the weather. A mild day can mean the possibility of recreation. If my chores and freelance work are done I grant myself permission to do things I love, such as horseback riding and archery practice. It's both a reward system and stress relief, a way to unwind from worries about bills and the mortgage. When my daily task list is done I can relax a little, and leave the property for a few hours and visit friends on neighboring farms. This is the gift for the self-employed, a bit of freedom. But like most swords, this claymore's got a double edge. Because when snow falls deep and the temperatures drop below the freezing point, it usually means that I am staying put. I could have finished my deskwork a week in advance, but no sleigh bells will be jingling on this farm—there simply isn't time. Severe weather takes this farmer's dance card and fills it.

In a storm, be it a late-summer hurricane (like the destructive Irene, which crashed through Vermont and New York in 2011) or winter nor'easter, those slower times of my day when there is a lull between the chores of milking, feeding, and hauling water in buckets from the well are now given extreme tasks. They take precedence over any sort of errands or bills, appointments or meetings. The day is packed with repairing blown fuses, shielding everyone from high winds, raking snow off the old slate rooftop, defrosting frozen water troughs by chopping through the ice with a hand ax or sledgehammer, chopping extra firewood, and keeping every living thing on the farm comfortable. I've grown used to having to earn my heat. It feels better, the heat from wood. The idea of a life that demands this sort of loyalty to home is almost lost in modern culture. To me, it is just Tuesday. Give me fire.

But the costs of wood heat's earthy blessings are paid for in proximity. Through much of the winter I can never be more than a few hundred yards away from the woodstove, which has a mighty hungry maw and requires the toughest feeding schedule of anything on this farm. I oblige. I keep the coals steady so that the house remains a place of comfort and respite, regardless of the weather outside. How important a warm someplace is in a January twilight after one has unloaded a truck full of square hay bales in freezing rain! That fire is therapy, inspiration, and a healing balm far more potent than anything on the pharmacy shelf. It's cold medicine and a shot of whisky, both—a little hope and a little force.

This scrappy place requires absolute presence. Being a homesteader, especially a single homesteader, is a lot like voluntary house arrest. Keeping dairy goats means that you don't just stay home for Christmas; you are housebound. But that dedication to place is exactly why I chose this life. The work of keeping the farm productive (and keeping yourself mentally functioning) is a maddening and intoxicating fistfight with voluntary circumstance.

I love the early mornings that get me outdoors before the sun crests the tree line. And I love the early nights tucked in under heavy blankets with my kind dogs, too tired and too grateful for their heat to kick them out from under the covers. I love holding baby chicks in my dirty palms and feeling their rapid-fire heartbeats under baby down. I love the black soil of spring, the sweat and humidity of summer, and even the downpours that wash away three months of work in the garden. To become a farmer is to accept the worst offerings of chance and laugh at them, understand that there is no difference between the pleasure or the pain because feeling either is proof you are still waltzing among the living. You pray to the fireflies and black flies alike. You learn to love both the smell of baby lambs and the blood buckets on hog-slaughtering day. They are the exact same thing in the end. Life. All of this, proof positive that life surrounds us, and leaves us, and we're just another bit of blood and meat blessed with pleasure.

Cold Antler Farm has always been a single woman's operation, with one employee, but that hasn't slowed me down. To the contrary, over the years Cold Antler has graduated from a handful of chickens and a few rabbits into my full-time job. I now raise dairy goats. I turn that milk into cheese and soap. I raise rabbits, pigs, and chickens for their meat. I keep hens for eggs. There are vegetable gardens and beehives, too. I use horses as working animals to cart, haul, and plow. There are no tractors on this mountain farm, just a strong brick house of a Scottish pony and my stubbornness. So the farm runs entirely on animal power (usually I am the animal powering it) without the aid of tractors. I'm not against tractors; I simply can't afford one. And even if I could, I am certain it would kill me on my steep hillsides right quick. Food is grown on a small scale, using hand tools, and every year I invest time in making a little more space in the garden. The garden is always a small-scale project when compared to the animals, but it keeps me in salad greens, tomatoes, onions, garlic, savory and medicinal herbs, squash, and potatoes and other root crops. I call it a jar garden. Everything from it can be canned into jars or turned into herbal tinctures or salves.

This is what takes up my daylight, and keeps me up in the darkness. It's a lot of things to me, but mostly love. I'm in a monogamous relationship with six and a half acres cut into a mountainside.

I find just as much drama and joy in seedlings and fresh basil atop a goat-cheese pizza as in the story of a lamb turning into a roast for a Yuletide feast. This is the work I do here. I create amazing meals and memories from the sordid work of dirt, death, and sex. Only it isn't sordid at all, not anymore. I was a devout vegetarian for ten years, and I am now a shepherd on horseback. The transformation that agriculture has put me through was sometimes chosen, sometimes thrust upon me, and sometimes completely unexpected. I didn't get into homesteading with any intentions of raising animals for food, and now it is what this farm is known for. I used to think nothing of taking a day off to wander through New York City museums and restaurants, and now, although I am just a three-hour train ride away, New York City seems as far away as the Louvre. Farming has changed nearly every concept I had about animals, travel, food, time well spent, and what hard work really is. It's altered my preferences and made me tougher in matters of the body and a lot softer in matters of the heart. Being a homesteader means accepting all of life, the beautiful and the horrible, the grimy and the passionate, as one blessed thing. It's all the same now. And I don't have to go out into the world seeking anymore. My life has led me to six and a half acres that have forced me, once again, to jump—into chance and dramatic change, and now the people, community, farmers, and landscape around me are my whole world.

To choose to live so locally, to become dependent on your own community and backyard, has gone out of fashion over the past hundred years or so. It's in aiming for this kind of closeness to place that I am finding my path to happiness. My goal is to create a farm that inspires me, feeds me, heals me, and teaches me. I want to know the work and effort of growing food and working beside animals so I don't forget that I am one, too. I live not by days of the week but seasons and the intense work required by each of them. Life isn't simple, but it is basic, and that is a relief I have sought since I first learned I could have a taste for it.

I see the craving for being rooted to home coming back, and strong. Not in the sense it has meant to most of us, meaning moving into your own home and starting a family, but in the sense of digging in wherever you find yourself. Becoming part of a community you can contribute to and finding all the needs of mind, body, and soul a horse cart ride from your front door. It wasn't all that long ago that people considered fifteen miles a day's travel. They may not have had as many stamps in their passports or stories about bars in Ireland or Shanghai, but they did know enough to prepare for winter by stocking up on food, firewood, and water. Why is it that today, a successful life means possessions and purchased experiences and not actually succeeding at sustaining your own life? Why are Indian vacations and handbags more sought after than learning to play the mandolin in our own backyard? I feel like we've been tricked. Convinced to want more than we need and to covet the approval of others the whole while.

This may be the most terrifying trap in the world to me, to live someone else's idea of a good life. I'm too short on time, as we all are, to live up to someone else's permission.

COLD CROPS

WHEN I WAS GROWING UP in the Lehigh Valley of Pennsylvania, no other cable channel was as important as the Weather Channel. The television in the kitchen was basically like a radar machine, constantly tuned to that one famous green screen of moving clouds. My mother was a public-school teacher, so inclement weather meant changes in her scheduling and days off in the winter. In the summer the channel was left on out of habit, and constantly knowing what was happening meteorologically helped us plan regular yard work and outings too. Some days were poolside days and other days were good for a trip into Allentown for errands or the mall.

Being from a family so vigilant about precipitation, I have memories of watching people and places on the Weather Channel and thinking about how wild it must be to live in those foreign worlds. I watched places without winters, such as San Diego, where it always seemed to be seventy degrees and sunny outside. I watched specials about tornadoes in the Midwest, hurricanes swirling up the southeastern coast, and ice storms in New England that left people without electricity for weeks. Forget cartoons and the *Baby-Sitter's Club*, this was entertainment for me.

One image from those days I cannot shake was a report on winters in Albany, New York. Video footage showed a man walking over the top of a STOP sign on his way down a road. I could not fathom that much snow. Pennsylvania got its share, but to see the plows pile it so high after a single storm that it would bury traffic signs was fall-down-the-stairs incredulous. Albany, New York, huh? In my mind, Albany became a tundra where brave men and women had dogsleds in their garages and ice-fished for sustenance when the circumstances demanded, which had to be regularly with snowfalls such as that. Who would choose to live in a place like that?

Me, it turns out.

My farm is an hour north of New York State's isolated, northern capital city. Although I have yet to climb over a STOP sign while walking downtown, winters here are no joke. They last a long time, too, with our first frost-free date practically in June. This makes gardening an act of faith and strategy in ways people in gentler climates have nightmares about.

The first things in the ground in the spring are cold crops, hardy beasts of vegetables that laugh in the face of cold fronts and barely flinch at a touch of frost. They may seem humble to the early tomatoes my friends in the South have, but after winters like ours, a radish shoot is reason enough to crack open a bottle of champagne. Respite comes in many forms, but few are as welcomed and needed by soul and skillet alike as the first new vegetables of spring.

My spring garden is tilled up with a hoe, shovel, and pitchfork as soon as I can work the ground (which is generally in April), and then I seed directly into the soil. My growing medium is a combination of the rocky topsoil and compost from the piles of animals' bedding and manure that break down over the seasons. Any particular garden bed may receive the gifts of last winter's pigs', chickens', and horses' plops. Time and the magical heat of decomposition turn the manure into a black, worm-filled powerhouse of veggie nutrition.

This is when I set up a trellis for peas, space out the kale, and have daydreams about more tropical adventures such as squash and tomatoes. I need to temper those lustful thoughts because right now the only color this little kitchen garden sees is green. Greens in the form of baby lettuces, crinkly kale, onion sprouts, broad-leafed broccoli, and twirling pea vines. These are the old standards, garden songs I know well and have come to love. If I wanted to I could drive to the grocery store and buy green vegetables shipped from across the country or imported via ocean liners and airplanes, but that kind of eating has started to feel disingenuous to me—too rich for my blood. Honestly, it also tastes like garbage, bland and wilted compared to the crisp, flavor-packed harvest from a spring garden. Nothing store-bought compares to the crunch of a snap pea picked right off the vine, or the taste of kale evening-harvested from its bed, when it's been flash-fried in olive oil and seasoned with herbs.

All winter I eat good, filling meals, but they mostly consist of root vegetables, meats, and soups. The kind of comfortable foods you mop up with a chunk of heavy bread, that leave you ready to curl up and take a nap by midafternoon. But as spring awakens I crave lighter fare; my body screams for it.

I'll admit, I am not a purist. I can be seen on occasion in the produce section of a local market loading Californian kale into my caddy, but it is never the same as kale from the garden. Travel like that makes any of us careworn, especially vegetables. I prefer to put in the sweat equity for my kitchen dreams. Once you go backyard you never really go back.

And when I am out there in the early spring sun, feeling a glow of sweat gather on my arms and those old gardening muscles stretch again, I no longer see my adopted state as a foreign land. It's a lot warmer than those sensationalized Weather Channel shorts show. And ironically, it's the place that has trained me to look forward to green things in ways my fourth-grade self could never have understood. Afternoons at the mall and mild winters are part of a past life, but my future is lined with good manure and trained vines bursting with white flowers and the promise of a solid Amish Snap in my salad.

I no longer watch the Weather Channel, but I imagine their big-fish tales have only grown larger. And if any Mid-Atlantic preteens are viewing Albany as a wild place, I hope they get the urge to discover its softer side. The salad is to die for.

LAMBS ON THE GROUND

UNLIKE OTHER SPRING PROJECTS, such as gardens and pigs, lambs do not confine their activity to a particular space, and their arrival at the farm isn't anywhere near as orderly as piglets delivered to the farm in pickup trucks and gently set in the barn. Lambs here at Cold Antler are not brought in as young animals birthed elsewhere and raised to the point of harvest, be it for wool or meat. Lambs at this farm come from my own ewes each spring. A ram services my ewes in the fall, sometime between October and December. I tend a mixed flock of half Scottish Blackface and half English Longwool breeds. The arrival of their little ones is the most stressful, wonderful, and exciting time this farm sees at any time of year.

I can plan their arrival with some foresight, but it isn't an exact science. The only semblance of control I have is to choose the time the breeding ram joins the girls. I do not keep a ram at Cold Antler full-time. To do so would be tricky and have one of two outcomes. Either I would have to house him separately, away from the ewes, which is stressful for any herd animal, or house him in a fraternity with some wethers (castrated rams). Many farmers decide to house the ram alone, but this has always rubbed me the wrong way. Keeping one sheep away from his flock can only cause the animal anxiety, and cost you a lot in separate fencing, housing, and more time feeding and watering another pen. And it can't be just any pen: that solo ram will not want to stay where he has been assigned. So ram pens need some serious panel fencing and electric wire. All that jail time for a poor fellow who just wants to treat his ladies right.

I could keep two rams, and let them trade off breeding duties every year. That way the rams would have companionship, but then I would still be confining them and creating extra work for myself. So instead of keeping a testosterone-loaded gun within firing range of many fertile females I do rent-a-ram.

For small-flock owners, renting a ram for fall breeding is a common practice. You pay another farmer who has more stock a fee, and he delivers a ram of your choice to your flock every season. For this farm that means right around early December. Along with thoughts of Holly Kings, mistletoe, and gingerbread cookies dancing in my head are also thoughts of sheep sex. That sounds sordid, but really, it's just business.

I depend on each year's lambs for much more than wool and roasted joints. Ewe lambs are a blessing; ewe lambs stay at my farm to replenish losses of older animals in recent years. Ram lambs are usually pre-sold to other new and local farmers looking for proven stock for their own smallholdings. What isn't pre-sold is almost always bartered for farm necessities such as firewood or hay futures. My life is very much like the board game Settlers of Catan.

For the past few years my ram, Atlas, has been shared between my farm in Washington County, New York, and a friend's farm in the Adirondacks. He spends most of this time up north with a herd of Scottish Highland Cattle and a few ewes, but every year he returns to Cold Antler, the place he grew up. Atlas comes "home" to my farm to breed in the fall. Since he was bought as a ram lamb and isn't related to any of my sheep, he is perfect for the job. It's become a tradition around here, the Return of the Ram. And it is one of the new holidays that have taken over my life.

When you live with sheep, and regardless of your religion, you become a member of the Society of Lamb and Wool—a shepherd. Taking the crook, even if only metaphorically, means understanding that you are now adopting a whole new calendar. It's the datebook of breeding, shearing, lambing, pasture rotation, flushing, harvest, and breeding again. I wish they made greeting cards I could mail out to friends and family to announce and celebrate such holy events. I want everyone in my life, farmer or not, to know the joys of "Naked Sheep Month!"—when shearing—and "It's five boys and four girls!" with cigars taped to the envelopes. There just aren't enough of us with lanolin under our fingernails to tempt greeting card companies, I guess (indie letterpress designers—I know you're listening). Every year these holidays of wool and rebirth bring me to a new place of reverence for the farm and the world I am blessed enough to live in.

And what could be a better symbol of blessing than a newborn lamb on green grass after a long winter? Those leaping and bleating babes melt away the last bits of winter clinging to the insides of my ribcage. I can have an entire cold frame bursting with chard and kale, green as God's eyes, but nothing compares to cradling a lamb in your arms.

The shepherd's kitchen is not a pretty sight in the springtime. The world outside may be flower-sprinkled and green, but inside the farmhouse you can expect buckets, bloody towels stained with placenta, bottles, ear-tagging guns, syringes, tail bands, and mud. The business of bringing lambs into the world is a messy one. For every adorable little fluff ball gamboling among the mountain thistle there is a very angry washing machine, and a very tired farmer. The exhaustion is from the constant vigilance as much as from any of the births. During lambing season you must check on your flock every few hours, and that includes through the night, in case one of your ewes goes into labor. And twins are common after a ewe's first pregnancy, leading to complications. It seems that my own flock likes to wait for the worst possible weather to roll in at three A.M. to hit hard labor. By the time the babes are cleaned up, tagged, their tails banded, and suckling I am ready to collapse into bed, only to realize another mama is about to burst.

So there is no rest. You soldier through the season, making sure you are present and able to help at any birth that may need you. Every night is a series of alarms going off, and you need to have the self-control not to just stay in bed. Once up, you change into the warmest clothing you can find, call your dog, and head out with a flashlight into the pasture. Snowing, raining, starlit, or blessedly mild, it doesn't matter—you walk out in lantern light or sunlight alike, eyes keen for signs of ewes ready for their midwife.

My third year of lambing, I had only half a dozen ewes giving birth. I felt prepared and confident in the work ahead. My supplies were ready, my body was rested, and I had five pounds of coffee ground and lying in wait in the freezer. I had a date circled on my calendar that indicated about when the first lamb of the season was to be born. "About" being the operative word, because unlike the attended and well-documented breeding of the dairy goats, sheep go rogue.

The date I had circled was exactly five months to the day from when the ram was introduced to the flock, back in late fall. Sheep come into heat every sixteen to seventeen days in the fall, generally staying in sync with one another, so when the first lamb is on the ground you can expect the rest to follow in short order. But it's that first lamb that's a toss-up. If the ewes were eager and ready to go upon meeting their gentleman caller, that early circled date may very well be the day I head outside with buckets and towels. It could also be a month and a half later. And there is no way to predict, plan, or guarantee a thing.

There are better ways to orchestrate breeding, and thus lambing. Shepherds can put a special harness on their breeding ram that is loaded with a sticky chalk on their chest. The reason being that every time the mister mounts a missus, he leaves a neon, blotchy tramp stamp right above the ewe's tail. If you write down the date each of these splotches appears on your ewes you can better estimate the day a little one will arrive. I tried this approach my second year lambing and the harness was too large for the little ram, so it slid right off. I decided to let nature take her course instead. Sheep had been popping out lambs long before I ever held a leather sex-chalk harness in my hands.

My own Scottish Blackfaces and English Longwools rarely need help with the actual delivering of their babes, but they do need care and assistance Mother Nature isn't prepared to provide. When the lambs are on the ground I am ready with clean towels to help clean and rub the babies into waking life. A just-born lamb is still coated in the birthing sac and needs its mouth, nose, eyes, and ears cleaned and opened to the world. A good mother does this, but many of my sheep are first-time mothers, and I don't want to chance losing a charge to suffocation because its mother forgot to read *What to Expect When You're Expecting*. Birth problems can happen, as in any pregnancy.

When the little ones are bleating and breathing on their own I pick them up and slowly carry them into a small pen set up as a nursery for the new parent. These little pens are called jugs by folks in the sheep world. Any sort of four-by-four pen will do so long as it has clean bedding, water, grain for the hungry mother, and room for the little lambs to lie down without being squished. I carry the slippery critter into the pen with the mom carrying on bleating the whole time. When they are both safely inside I pour molasses into the mother's drinking water and watch to make sure the lamb feeds. Sometimes this takes moments and sometimes it doesn't happen at all. In those cases I have to either milk the sheep and bottle-feed that essential colostrum or hold the mother in place against a barn wall while showing its lamb where to find the teat.

Once the mother and her lamb (or lambs, as twins are common) are together in their jug, the baby feeding, and mother is set up with sugar-enhanced water and sweet feed, I let them be for a while before returning later in the day with the tail-docking band, ear tags, lamb paste nutritional supplement, and tetanus shot. If all goes as planned, that is the extent of the work the lambs will need from me for the weaning season.

To some, docking the tails is a barbaric act, but it is necessary for a healthy animal. I put a tight rubber band around the tails a few inches from their bodies. The tails fall off naturally from having the circulation cut off. Sheep are born with long tails, like Labradors. The tails are as thick as otters' and are bony and adorable when they are wagging like a puppy's. But sheep grow wool on those tails along with the rest of their bodies, and if a case of diarrhea or illness occurs, they quickly get coated with sick. This attracts flies, who ruthlessly bite, burrow, lay eggs, and eat the flesh off the lambs, a disease called fly strike that is not something I suggest looking up on the Internet if you value keeping your lunch down.

So with the babes' tails docked and their bodies equipped to fight off any infection, their mothers will be in charge of their care and feeding from then on. This is what I love about life as a sheep rearer: the postseason. You go through the messy spring of removing placenta stains from your jeans and living off coffee and quick meals, but once the real heat of a May afternoon arrives, you have green grass and a healthy flock teeming with new life and possibility. A month after the chaos of lambing, I have forgotten the sleepwalking and see the joy of it all. Before me are future skeins of yarn, knitted hats and gloves, lamb chops and crown rib roasts, real fleece blankets, and herding lessons with Gibson—the projects and sustenance of this farm and my community for months to come. What animal can offer you so much as a fine sheep? You get the promise of the product, sure, but that pales in comparison to the sense of goodness offered by animals that keep you alive, warm, and proud.

A RHAPSODY OF BASIL

THE SPRING GARDEN is a place of high performance art. No hot yoga class will make you stretch or sweat more. No workout video will train your arm muscles like a hoe and pickax. Your body changes in the work. Winter fat fades away, a combination of time outside sweating in the hot spring sun and the hours you forget to eat while lost in the hoary meditation that is ripping out burdock sprouts. You crouch down like a kung fu master, crabwalking along the rows, weeding.

Spring begins in the dirt. As soon as the garden soil defrosts enough to work the soil with a spade and hoe I am out there; my hands are starving for it. Working the first spring soil is like starting up a car with fresh fuel after being stranded for days. It's not so much the effort as it is the relief. In the winter I miss that feeling in my shoulders that I get when a bed has been awakened and given its first injection of new life. In my case, new life comes in the form of composted horse and pig manure, and old bedding hay from pens and coops. The melody of animal waste and decomposing carbon in the form of old hay and straw is a song written by time and death. It isn't pretty. Compost is always a hymn, in Dorian mode, like an Appalachian murder ballad. The fresh food of new compost is worked and turned into the old ground, and earlier (always earlier) than I should I start planting the first greens of spring. I sow

48

directly into the ground the seeds of lettuces, spinach, kale, and snap peas. I used to plant potatoes, but I learned that a late frost kills the leaves, and even if they manage through some act of holy will to make it through the summer, the potato bugs and blights of a wet May nearly always rot them where they are planted in the ground. The old-timers here in Washington County say to plant potatoes late, in June. Summer planting seems like I'm sleeping in, but as they say, "They are just as ready in the fall" and don't rot or become bug feast. I do as I am told when it comes to gardening.

Animals make sense to me. Their care, language, and life cycles have become woven into mine very naturally. Living with my dairy, meat, egg, and wool stock was not something I would call work, just a lifestyle choice. But gardens? Gardens are work, a second language spoken far from my native land. Their life and needs are so simple compared to those of sheep and horses that it is hard for me to get to that calm place of reason and chemistry to keep them content. You can muscle a ram into a shearing position. You can change feed rations, light sources, and nesting boxes to amp up chickens' egg production, but plants are not something you can manipulate through force or circumstances. Not really. Because problems are not solved through simple actions with plants, it's not that kind of game. If you have a lettuce crop that is bolting or (gasp) burning to a crisp in too much direct sunlight, you can't just dig it up and move it and expect it to perfectly adapt its roots to new soil. It might adapt and grow, but more commonly in my experience it withers from the trauma and slugs eat it. So what now? Set up shade umbrellas and deep-mulch the soil around the lettuce heads? That may solve one problem but it may be too much shade for nearby peppers or tomatoes that get an unwelcomed bit of melancholy come three P.M. So now you are trying to remember when to adjust umbrellas at what time and place, and even if you leap to that sort of idiocy you may not have realized that that mulch you brought in invited a whole new variety of slug egg into your garden, twice as voracious as any your little raised beds have ever seen. Gods alive, if a pig gave me this much grief I'd be driven to hang up my hat and wash my hands of it. But vegetables don't play by those rules. And so every year you learn from mistakes, over and over, and hope somehow you figure out the best way to not kill things next time around.

So, gardening isn't my strong point. That doesn't mean I don't love it. And it certainly doesn't mean I don't crave it with a deep ache come April's sunshine. When the first truly warm day arrives all that cynicism leaves me. I am completely entranced by my vegetable prospects, and while I do not lose sight of the animals and their role on my little freehold, I am having a blatant affair with seeds and soil. And what starts out as blind passion, that almost primal need to be out in the muck with tools and seeds, warms into a mature relationship come June. When I can take off my shoes and stand in the garden barefoot weeding, and feel the sun start to cook my skin—I'm a goner. All that labor out in the overexposed spring light, all those sore shoulders and mornings when it is hard to lift my arms up high enough to put on a sweater—those days are forgotten. I'm not sure how a few rows of plants fool me with this magic trick every year, but they do. If you let yourself love gardening, you become its willing subject. It's a happy kind of feudalism, fueled with really good pesto instead of gruel.

What grows in early spring is not the sexy vegetables of midsummer. If you were to glance outside my kitchen window in April you would see a lot of green, and the occasional white or yellow blossom, but even they remain demure compared to the voluptuousness incarnate that are tomatoes and squash. No, the big fruits and flowers are yet weeks away. Instead there are rows of salad and cooking greens, broccoli, peas, carrot tops, and other homely bits. I plant some radishes, too, only a few. But even though the early garden is fairly limited in palate, it holds the promise of some fresh greens to side with winter's diet of meat and potatoes and canned goods. It is easy to put in the extra time the garden demands when you are recalling memories of late-summer meals of Russian kale cooked under a Cornish hen, and the combination of wonderful bittersweet forest tastes with a coating of white juices from the small bird. That first little farm-raised bird roasted on a bed of swirling kale leaves feels like a holiday, a midsummer festival of gratitude and hours banked outside.

But Cornish hens and plates of kale are still weeks away. All the young, petulant garden has to offer is kale shoots and clumps of green lettuce. I use a chunky short stick, about the same girth as a paper-towel tube, and start making indentations into the dark brown soil just five to seven inches deep, near the roots of the new guys. Into each of these little caves I insert a scoop of the black-as-sin vermi-compost from my kitchen worm bin. This is what happens to kitchen scraps, given time and a couple of hundred red wigglers. It is so rich it feels vulgar in my hands, the Belgian dark chocolate of my garden arsenal.

And while I feed my new shoots I think about the seeds in my pocket. The seeds that make a backyard garden worth every day of effort: basil. This leafy, aromatic, blessed plant may be the one reason I cannot go without a garden in my backyard. Other people in this community can grow better tomatoes, larger pumpkins, and healthier snap peas. But it is the luxury of walking outside to my garden with a steaming slice of homemade pizza just out of the oven and placing some just-picked evening basil on it that keeps me planting. Stronger than any narcotic, more addictive than any drug, fresh basil isn't just about the taste but about the rawness. Such a delicate leaf, on a plant meant for milder climates than mine. When I pluck it from the stem and place it on some fresh, lightly toasted mozzarella, it doesn't explode in flavor as much as change the entire climate of taste in my mouth. It's rich without being overbearing. It complements the tomato and cheese perfectly, the missing link in a necessary trio of summer rites.

From seed catalogue selection on frigid January nights to planting in May to sunburn and sore arms to the first taste in late June—that isn't just an herb, it's a saga. And as I plant those first little seeds into the ground I do so with a prayer and a stomach rumbling with a hunger that will not be sated for months.

It is always worth the wait.

Gardener's Pesto

I use pesto for a lot of things—it can wait in the fridge for days while you decide the best way to use it. It is delicious on pasta, of course. For a quick summer meal, boil up some noodles and top with a dollop of pesto. It is also a wonderful dip for homemade breads. It is a fine salad dressing. I love to eat it with sun-warm garden tomatoes sliced in half and sprinkled with salt and sugar—it's summer on a plate.

My favorite pesto recipe is incredibly easy. All you need is two cups of basil leaves (packed well into a measuring cup or jar to measure), some olive oil, garlic cloves, and salt and pepper. You can also add other greens such as fresh spinach, chard, kale, parsley and rocket, but make sure that at least half of the green mash is that beloved basil. It really is the heart of pesto to this girl. Get out the food processor and you are practically done.

2 CUPS BASIL OR OTHER GREENS
⅔ CUP OLIVE OIL
2 CLOVES GARLIC
SALT AND PEPPER TO TASTE

Rinse and air-dry the leaves. Put all ingredients in the bowl of a food processor and pulse until you have a paste. Now you are basically there. All that is left to do is add salt and pepper to taste. But you can also spice it up a bit if you like. Some folks sprinkle soft goat cheese into their pesto or add pine nuts or other nuts during the blending process. At this point it is ready to serve or store in the freezer in plastic containers or storage bags. When I am whipping up a batch of fresh pesto I rarely add pine nuts or Parmesan . . . mostly because I never have them or make them around here!

OSTARA

THE SPRING EQUINOX, known around here as Ostara, is a big deal at this farm. It's a full-blown holiday, just as important as any modern family's Easter or Ramadan. On this little farm it's a celebration of coming life, love, work, and luck. It's the awakening of the entire season of work, birth, growth, and eventually (fingers crossed) harvest. Plant and animal life are both seeded with the intention to consume them later in the year. It's not a time of focusing on abundance, like summer and fall, but more on that ecstatic tension a sprinter feels before a race. A whole lot of energy is about to explode over the next few months, and Ostara is the starting line.

Named for Ēostre, an ancient fertility goddess honored by Germanic tribes, Ostara is a modern re-creation of the old festivals of fertility and of a sleeping earth coming back to life. (It is also the root of the word *Easter.*) Think about what it meant in a culture without grocery stores or canned goods to start tilling the soil and feeling mud between your toes after a long, frigid winter. How could you feel anything but gratitude? While the experience for me is nowhere near the primal levels of necessity those earlier people felt, I too know what it means to feel blessed and relieved at mud puddles and to wave good-bye to the last pile of melting snow. Spring is a party stirring up new life and the earth itself.

Lambs, chicks, bunnies, turkey poults, and goat kids are every-where. Piglets are purchased and delivered in early summer, and seedlings spring to life in cold frames or out of the sturdy, hardy ground. I find myself addicted to the warmer weather, staying out-side longer and working harder than I did in the winter, and it makes for a better night's rest (especially after weeks of getting up every few hours to check on goat births or lambing pens). The barn becomes a place of such life, it is vibrating from every creaking and fading board.

When my goats' kids are born and I begin the twice-daily milk-ing, I notice the honey bees every so often float down from their hive in the barn's roof. I am, literally, living in the land of milk and honey. A place I know from stories of my childhood, but also a place I know in my own backyard. This connection to myth through food is why days like Ostara have stuck all this time.

Myth is a central part of my life and my spirituality. The idea that a story can last long enough to define a culture reminds me of farming; so very alike in the way both myths and farming sustain the human story. We need inspiration, hope, and tradition as much as we need calories. Together they feed the whole person.

Some people find inspiration in favorite folktales or legends much as the Celts did with their Tuatha Dé Danann and the Vi-kings with their sagas. For others, the larger-than-life heroes of recorded history have become their mythical heroes over the cen-turies. A few people might get their inspiration from beloved epic novels or epic movies. For some people inspiration has nothing to do with entertainment but instead with their faith, with the rock-solid beliefs that create this foundation of story and song. Be it tall tales or Scripture, humankind loves a good story.

The pre-Christian holidays of the Anglo-Saxons are the holy

days I keep. The old Wheel of the Year is not a symbolic series of holidays—here the observance of special days is practiced literally through the farm's own cycles and seasons. As a farmer I am experiencing the same fears, joys, and gratitude as did those who came thousands of years before me. That connection to ancient ancestors means so very much to me, to look back and see how not alone I am in my struggles and laughter.

Farming has a longer history than most professions. When you take the time to study the first agricultural cultures you quickly realize that their entire life—work, religion, celebration, and sorrow—was focused around food. Before markets and refrigerators—hell, before smoking meat was invented—early native people created a life around food: hunting, gathering, shelter, and survival. They devised the first holidays based on the solar year. These were the foundations of the modern celebrations we see printed on greeting cards today.

I don't see a lot of difference between the idolization of Abraham in the Old Testament and Abraham Lincoln. We just want to look up to what came before us. It means something to know we are the heirs of such persons, because maybe a spark in us would be capable of living such a life as theirs. Myth, viewed literally, is obscure, and we still carry vaguely historical events close to our hearts. They are as important to me as food and water. They are the reason I farm—I want to keep this story going.

DELIVERY

YEARS AGO WHEN I WAS merely dabbling with the idea of raising livestock, it was chickens that handed me my initiate's guide-
book. I worked forty hours a week as a graphic designer, and these
birds seemed like the perfect level of commitment. A whole flock
was less work than a pet parrot, and instead of incessantly repeating
my cell phone ringtone, these birds provided omelets. Score.

It is in the in-between time of late winter and early spring that
I am most fowl-minded. Hatchery catalogues, like seed catalogues,
litter coffee tables and bedside nightstands. It's a different sort of
perusing though, not the lustful way I page through Baker Creek
Seeds or Johnny's catalogues. This is a mission, and the result is
more than just eggs and broilers. Chickens are the core of this little
farm.

I had lived with working dogs and rat-catching cats before,
but chickens led to a different relationship. Unlike cats and dogs
there was no mistaking the colorful, fat hens in my backyard for
pets. They were "farm animals" and they started a paradigm shift
that allowed me to redefine my backyard. Backyards had always
been places of cut grass and leisure. My mother and father spent
hours mowing the lawn, planting ornamental flowers, and hosting
parties on the deck. A backyard was a place you treated like roy-
alty, a privilege to have and to maintain, and a respite from inside

work. Lounge chairs, iced tea, playing in sprinklers—this was the backyard I knew. Our pet cats and dogs played there with us, too. Wild animals like squirrels and rabbits and songbirds also enjoyed it, but they were more scenery than coinhabitants. Years later and thousands of miles from my childhood backyard, when I decided to turn my lawn into a place of food production, this world was turned on its head. It wasn't an act of rebellion, nor an act of destruction. I was simply changing how I was going to tend to this, my allotted bit of earth.

I set up a little coop and got some chickens. The grassy area around my home was no longer a place that just took things from the world; it gave back. It felt liberating. I fell in love with turning what I kept into something that kept me alive. I would never go back to pink "annuals," not unless the bloom was on the top of a potato plant.

Since purchasing my first five Black Silkie Bantam chicks in Idaho no April has passed without my recurring order of birds. "Box" is the correct term, too: They are delivered to your local post office in a special rectangular cardboard chick-transportation device divided into little compartments with air holes. The birds are packed tightly enough that their body heat keeps them warm, and the container is shallow enough to prevent them from climbing on top of each other and smothering each other.

The first forty-eight hours of a chick's life are fairly inconsequential. The birds are filled to bursting with the goodies inside the egg they just hatched from. They are not hungry, and they are ready to be moved quickly to their new homes. So as the fluffballs dry from their egg, brand-new to the world, they are rounded up, sexed, and mailed to farmers like me who ordered them in advance.

If this sounds a little cruel, please let me explain. As much as we like to anthropomorphize animals, in these first days of life those chicks are not akin to people being crammed into boxcars. They are only thinking about warmth and food—both of which they have in plenty on that ride of their lives. And where these chicks arrive is the important part of the story. The small mail-order companies that send chicks through the post allow small, organic, sustainable farmers to raise the alternative to the factory-farmed egg or broiler. The large meat and egg companies have their own hatcheries, while places like Murray McMurray and the Freedom Ranger Hatchery are providing ethical alternatives to the grocery-store standard of shrink-wrapped carcasses and factory-farmed eggs.

Chicken pickup day is a holiday on my calendar. Birds that were ordered weeks before have been hatched, packaged, and shipped from a hatchery in Pennsylvania just for me. They arrive early, very early, and someone from the post office always calls well before sunrise to let me know a very loud box is waiting for me to pick up. In late March or early April, it is still cold enough here that a few inches of snow coat the fields. Before I have had a sip of coffee or brushed my teeth I heat up the cab of the truck and drive a few miles into town to the waiting area outside the loading dock of the Cambridge Post Office. I knock on a closed Dutch door, and like in a scene from *The Wizard of Oz*, a small man with a beard asks what I am doing here. I explain that I'm here for the chickens. He smiles and nods and breaks his stoic composure and I am handed a box of the most adorable babies. We head home to their heated brooder set up in my mud room, and before the hour is up they are scratching at clean wood chips under a heat lamp enjoying their first sips of water and bites of bird kibble.

Every chicken at this farm starts its life inside the house. The endearing chirps of the little birds can be heard through the wall while I am doing dishes or cooking dinner. Just the sound of them reminds me of what lies ahead, how much life will erupt on this little piece of land. Chicks come first, but soon seedlings will be sprouting out of the ground and snap pea vines will be climbing up the windowsills. Lambs, kids, and a pair of piglets are on the way as well. By mid-May I will be a different person, out of my winter hibernation and acclimated to eighteen-hour farm days and little sleep from night-watching the pregnant sheep and greeting the blessedly warm sunrises. These very chicks whose tiny voices can be heard in the background are an avatar of a life lived the best way I know how.

In a few weeks the chicks will have proper feathers, and be ready to move outside when the night temperatures remain above freezing. Good farm birds start out slow; it takes weeks and time and tender care to raise them. Chickens were the first critters who turned my head away from being a consumer and toward the adventures of being a producer. For this gift I thank them, and I happily do the work to welcome them to their new world, one wire-cage trip at a time until they are ready to take on the farm for themselves.

As I work on starting a new raised bed in my garden I set the cage down on the grass or in the pile of just-turned earth. They coo and try to turn their peeps into clucks as they scratch through the wire into their first taste of dirt. They scramble around the same large cage, pecking at dirt and digging for bugs. With the sun on their backs, the farmyard finally in view, you see how these little lives will bloom before you.

CODE OF THE PIG

WHILE THOSE FIRST SHOOTS of spring greens are just break-ing ground, there is an entirely different kind of new life making a ruckus on the farm. Inside the small hundred-year-old barn behind the farmhouse is a snug little pen that houses some of the most rewarding animals I have ever raised. They are brawlers and love bugs, are clever and coy. They are gluttons and thieves. They are proud and jealous. They are quick as cats and as foxy as foxes. They have mastered sins we humans would never dare to try. They are swine, and don't you dare underestimate them. Because when they aren't owning their smiles like a paid-off mortgage or perfecting the art of loafing about, they are plotting their escape.

I never planned to keep pigs. But now a farm without them seems incomplete. I started with a single little girl I bought one beautiful October day from a nearby dairy farm. The farm had a few litters of piglets a year as a small source of supplemental in-come. I brought the little pink Yorkshire home in a dog crate lined with straw and named her Pig. She had the pen all to herself, and she lived well in the barn through a very cold winter. I treated my first pig like a queen. She had the best food I could afford, a heat lamp, and enough bedding straw and spilled water buckets to cre-ate her own plaster company.

I now raise pigs in pairs, and I name my pigs, even though they will become food. Some people cringe at that idea, but I do not. I never invite a pig to the farm without knowing from day one that the little guy is going to be eaten by someone in my community or me. I accept their fate, and since I need to hang out with them several times a day I find that names happen even if you don't intend them to. After a while of saying "spotted pig" and "pink pig" it becomes Spotted and Pink, and suddenly they have monikers, intentions to keep them generic creatures lost. So I try to have fun with it, naming them something clever that works in pairs. One winter I had a pair named Kevin and Bacon. Lunchbox and Thermos. Rye and Whiskey, named after the spirits I savor as well as the old-time song "Rye Whisky" that I love playing on my fiddle,

It's a whiskey, you villain, you've been my downfall
You've kicked me, you've cuffed me, but I love you for all.

Each one of the swine that has graced this farm has had its own personality and gusto, its own way of being a pig. And watching a pair of piglets spooning together under the hay in the predawn barn might be one reason I remain a homesteader. Within their life story is the saga of the small farmer. The spunky verve of the curious piglets, animals that warm your heart and make you laugh just by sitting in their own feed dish and smiling. The speed of growth, health, and vigor of the hogs as they become adults, the moments of pride and success for the farmer as she watches her charges thrive and grow. The drama of the harvesting, from slaughter to butchering—which elicits a complex feeling of sacrifice and gratitude. And the final joys of meals shared with people you love from an animal you knew well. The lives of other animals raised for the table on a farm follow the same plot, but the pigs do it loudest.

Piglets arrived here at just about twelve weeks old, but unlike puppies or kittens of the same age they are remarkably matured. Each weighs close to thirty pounds (at least four times its birth weight). In a few days they will associate me with food and water; they will know that when I walk into the barn with a bucket they are about to become content beyond measure. The bucket's buffet will change with the seasons, but they can bet on being served the previous day's kitchen and garden scraps. Everything from eggshells to soft fruits to leftover barbeque sauce and slightly sour milk may be inside the meal bucket. The scoffers squirm and squeal at the sight of it—a pure happiness I never get tired of inspiring. It will take a little maneuvering and skill to dump equal quantities of the manna into their separate bowls of pig chow. You need to feed the little ones separately or you'll end up with one fat pig and one lanky wallflower.

Rye and Whiskey were brought in a beat-up farm truck by two fellows who looked as thin and hardworking as marathon runners. They wore old baseball hats, needed a shave, and had well-worn shirts on, but they smiled brightly when they greeted me. In the back of their old Toyota's cab was a little dog crate, and inside it were the girls I had been waiting for. I asked permission to open the cab's door and then did so. Two plump little gilts (young sows that have not farrowed) sniffed through the metal crate's bars. The truck and drivers may have been a little hardened by farm life, but these two were as pink and precious as could be. They had clear, alert eyes of a grayish hazel and clean skin and were already as plump as tuffets. They were a bit larger than I was expecting, but piglets always are. You expect something the size of a puppy, an animal you can cradle in your arms, but pigs leave that stage somewhere around five weeks old and certainly well past weaning. These two were off the bottle and loving life, ready to eat up whatever I could offer, and soon, please. They stood up and started pushing their shoulders into the cage door, squealing, and I thought it would be best to ask for some help getting them into their pen in the barn. It wasn't a far walk, exactly, but carrying fifty pounds of yelling animal in a metal cage would be a two-person job. I was ready for it, with sweat-stained work gloves on and my chore jeans ready, but the two gentlemen had other plans. Why bring the cage to the pen if you could just carry the pigs themselves?

I trusted them to pull this off, but not myself. Ten-week-old shoats may not seem like a hassle, but they are mostly muscle and force, strong and agile enough to twist out of any grip I could attempt. I know because I have lost more pigs than caught them while trying this method. And when you finally do catch the screaming little pink bullets the second time, they carry on in such a human, alerting bawl that any sane neighbor would call the police. But these men were no pig-slippers, no sir. These guys opened that crate door and in one adept motion caught the pigs by a front and rear hoof and carried them to the barn. I led the way with Gibson beside me. His eyes were huge at the sounds and waving fury of the little beasts in the calm men's grip. It felt like a parade.

At the barn it was quick work getting them into the pen. It was lined with fresh straw bedding. The pig pen shares a wall with the goat's pen, and both of my dairy maids popped their heads over the wall to take a look at their new flatmates. The little pigs froze in their new place. Their whole world had changed in a matter of hours, and now they were away from most of their siblings in a strange place with goats staring at them. They didn't bother with a drink from the clean bucket of well water, and they didn't even so much as sniff at the pan of piglet chow I had offered. They just stared up at us, the goats, and the black-and-white dog watching with open-mouthed panting and his tigerlike twitching tail. I couldn't blame them for their discomfort or lack of performance. If you manhandled me into a box I'd be likely to succumb to a little stage fright as well. We made sure their door was latched tight and I paid the men for the pigs and their time delivering them. They thanked me, wished me luck, and drove off. I stood in the front lawn with Gibson and let out a little sigh of relief. The hardest part of raising pigs, for me anyway, is the tricky work of transporting them. It's getting them in and out of their pen without their escaping that breeds anxiety, not the daily chores and feeding. Gibson watched the truck make the turn down the mountain that sent the men out of sight and then turned to trot back to the barn. He had pigs to keep an eye on now. So did I.

We walked back to the pen to check on the little girls, and were happy to see them walking slowly around and checking out their new quarters. They were quiet now, save for a few exploratory grunts. I poured a little cracked corn over their chicken feed and the action perked their interest. One of the little gals gave it a try and started crunching on a few kernels and that tripwired the Code of Pig in the other. As a rule of pigitude, when one pig is eating something, all other pigs should be, too. So they both dove into the feed and I leaned over the pen's gate to watch them do what they do best. Two happy pigs in a clean pen with food, water, hay to burrow and nest in, and attractive neighbors. What else could they possibly ask for?

These little pigs will grow in the fashion of meat hogs, which is long and fast. The rate at which these animals gain weight is alarming, and takes some getting used to. In the time it takes a kid or lamb to grow to half its adult size a pig can go from 4 to 250 pounds! They'll live at least another four months, and in that time they will eat their way from the size of a cocker spaniel to a high school wrestler, 190 to 225 pounds. Right now they are working on growing body length, which makes them seem cartoonishly dispro-portionate. They remind me of dachshunds at this stage, all stumpy legs and torso. As the months come on they'll get taller, and longer, and when they hit a certain height they go from shoat to hog. The weight starts to show in their jowls first, as they grow rounder faces and their eyes squint back into their expressions a bit. Then the belly gets round and the butt gets huge and from there on you have yourself some fine hoof stock. Before you know it you're setting up a date with the traveling butcher, a man who is busy here in farm country and requires appointments made months in advance. But butchers and bacon aren't on my mind now. Spring is a time when piglets get to just be piglets. It's happy work.

Every time I feed piglets it is with a bittersweet grin. It's fun as hell to watch them eat, but you can practically see the dollar signs being masticated between the carrot tops and pig chow. Few people who do not raise livestock understand the cost of raising feeder pigs like these. The pigs themselves are usually the smallest portion of the investment, ranging from fifty to a hundred dollars an animal around here. It is in the feeding and harvesting costs of slaughtering, butchering, smoking, and wrapping the meat where it gets expensive. A pair of hogs costs me over $440 to be slaugh-tered by a traveling team of kind workers here at Cold Antler, and to then be hung, butchered, and turned into products like bacon and hams. Add in the cost of food, even when it is supplemented by garden and house scraps, you are looking at easily $500 a pig in just upfront costs before a slice of bacon ever passes your hungry lips. And economics is the main reason I raise two: I raise one to sell and one (or part of one) to keep as meat for my own meals. People in my community buy a live share of a pig; they are co-owners of a living animal and will receive the meat come butchering time. It is my job as a farmer to use their money to buy the piglets, and their feed, and to cover slaughter and butchering costs at a share price that covers my time as well as my share of pork, which usually ends up being half a pig in the freezer. I do not sell pork; I give it to the shareholders. It's my thank-you for their making it possible to afford to raise pigs here on a writer's income.

This final stage of the saga keeps me raising pigs. Of all the farm dinners I have shared here, the ones the pigs offer satisfy the most. I can place a pork shoulder in the slow cooker overnight with a bottle of hard cider, some honey, barbecue sauce, and some brown sugar and in the morning the house will smell rich enough that you could spread the air over bread instead of butter. Pork-themed meals like this have made hungry guests salivate on cold nights. It has filled musicians' bellies in front of bonfires, topped summer garden BLTs with savor, and taught me to love lard as it used to be loved—as a conduit to perfect pie crust and fried chicken. There is no way to euphemize my pigs' role on this farm, which is to feed me and the people I love. They are celebrated until the freezer is sadly picked clean, hopefully just a few weeks before it is refilled by the new harvest.

SHEEP SCHOOL

LAMBS REMIND ME OF MYSELF in kindergarten: new to social interaction but eager to make friends. The little babes that were
nothing but cotton balls on stilts a few months ago have filled out
and grown brave. They run around in little packs, causing a ruckus.
But they don't start out that way. They start out, well, sheepish, for
lack of a better word.

For the first week or so after their birth they rarely leave their
mother's side, doing little more than sleeping and suckling. It's a
quiet time, being a new lamb in this world. You may catch twins
roughhousing a little here and there, but even that is rare around
these parts. Newborns are timid, followers of mama, and unless
trained on a bottle make themselves scarce when we heavy-footed
farmers tromp around. I have learned that if you don't catch a lamb
in the first thirty-six hours of its life you probably aren't going to
catch it at all, not without getting ready for a serious workout and
some comedic B-roll footage. And catching is important, as the
little ones need their tetanus shots, their tails docked, body condition inspected, ears tagged, and some helpful vitamin paste. If you
gather them up calmly, close to their mama, they are calm in your
arms and don't flinch or complain. I have gotten used to the tools
and motions of the tail bander and syringes; I have the tiny rubber bands in my pocket and have memorized the doses of vitamin

paste. The smell of new lambs has that same kind of "newness" as a new car. Not something you can pinpoint, but as certain as the texture of a brick. I hold them, breathe in their scent, and get to know these new creatures through my pediatric care. It's adorable work but somebody's got to do it.

Weeks pass, and the small ones start to learn what it takes to be a sheep, all the tricks of the trade. Here on the mountain they watch their mothers eat grass and ever so daintily give the green stuff a try. I can only imagine the contrast from warm milk to green strands. Whether it is stubbornness or peer pressure, they continue to nibble at it. Same goes for any hay that may be fed out between pasture rotations. (My land needs time to rest between feedings, and every few weeks it goes from open pasture to recovery time, and the flock switches to a diet of second-cut hay in a different paddock.) I love watching the woolies grip and pull at the long stalks of hay, complete with the occasional flower or twig, chewing as they look around at the rest of the flock.

Besides adopting the basic rules of diet and playing tag, the new lambs learn the easiest paths to the water trough and the best way to avoid muddy spots and sneak under the fence for a forest gorge-fest. A lamb's savvy differs from animal to animal and is a direct inheritance from its parents. I have a ewe named Ruckus, an older gal, who has learned every way to get under, around, over, or through a weak fence there is to exploit. Her lambs are just as clever. Yet another solid, younger sheep named Brick rarely wanders away from the fence line, and her little ones don't seem to try to sneak out until into their adolescence, when one of the "bad kids" shows them how.

This also reminds me of myself in kindergarten. I didn't dare color outside the lines until someone showed me I could and there weren't any real consequences when I did. It didn't take long to figure out I could flip the coloring sheet over and draw on the back, too. What it offered was a blank slate to create the world I wanted to be in, not feeling restrained by the lines and borders an adult gave me. I try to remember that when I watch a little ewe lamb squirm out of the sheep paddock and make a beeline for my lettuce patch. Like me she'd rather ask forgiveness than permission. A bite of speckled trout greens on a sunny day sure is worth being fussed around by Gibson, a sheepdog that doesn't even bite.

THE LIFE OF HAY

I LEANED OVER THE ALUMINUM GATE and felt the soft curve of the weathered metal tuck into my forearms. I don't know what it is about eight-foot metal gates, but they seem to be ergonomically engineered for leaning on. This could be due to their strength and smoothness, or possibly because most metal gates are looking over pastures and paddocks where beloved critters reside. I just can't help it. You see a sturdy (nonelectrified) place to ponder or converse and I'm drawn to it. I lift my sturdy five-foot-three-inch frame onto it and rest my weight on top of the four-foot-high gates, usually stepping a boot up on a lower rung. If you happened to drive by and wave you would not doubt for a second I was *reckoning* something, or about to dive into a good half hour of racing conversation about farm insurance or the myriad uses of baling twine. Now, I'm not the best gate leaner in the county, but I am learning fast. And one Friday morning, deep into the maw of April, I was learning from one of the best.

I was picking up a truckload of hay from Nelson Greene, the finest hay producer in the area, as far as I'm concerned. Now in his late seventies he is usually under the weather from various respiratory problems he's earned from a lifetime of living among cattle, hay chaff, dust, and silage. Nelson isn't in the pink of health, but he is far healthier than most folks his age that I meet. He is well over six feet tall and easily two hundred and forty pounds of worn-in farm muscle. He has hands that could palm a basketball and the ability to easily lift up a sixty-pound bale in each hand and load it onto my truck. One time he helped me get off a hay elevator by grabbing my belt, lifting me up in the air, and lowering me onto the ground. I was so shocked at his strength, I just stared at him. I weigh substantially more than a sixty-pound bale of hay, and he moved me aside as if I were one of his barn cats. I want to be like him when I grow up.

I have been buying hay from Nelson since the first day I had a serious need for it. He never lets me forget the story either. This was back when I first moved to Vermont from Idaho and had just acquired my first three sheep from a homeschooling family in Hebron. The day I picked up my sheep—and was transporting them back to my farmstead in the backseat of my Subaru—I crested a hill and saw a man piling beautiful green hay bales into a large red wagon being pulled by an ancient-looking red tractor. I pulled up to the work crew, rolled down my windows, and leaned over the passenger seat to ask in a loud yell, "Hey! Do you have any hay for sale?" to which Nelson looked down at me and the vehicular situation and replied, "Hay? YUP! Got any sheep?" I laughed and pulled over to strike a deal. I didn't know then but I had picked up my sheep right in the thick of second-cutting season at Nelson's farm, and if there was one thing he had in spades it was good green hay and plenty of it.

That very same evening I unloaded the sheep into their brand-new pen, and before I even got to lean on my own metal gate, I was back into that Subaru wagon and headed out to pick up my first load of hay. Nelson and I fit six bales into my car, and I felt as proud as I was the day I was handed my college diploma. It sounds like a ridiculous comparison, but I assure you it's accurate. It took four years to learn the skills and pass the exams required by the state of Pennsylvania to earn a bachelor of the arts, but it took twenty-seven years to get up the nerve to buy animals with hooves and then use a station wagon as a pickup truck.

Hay was a language I needed to learn. It's more than winter feed; it's a lifestyle choice. You can't turn back a life with hay once you start buying bales, because buying bales means you have made a serious commitment to how you spend your time. Buying hay means you are feeding at least one large herbivore, likely several, that needs it to sustain itself over the months when the grass is long dead and buried under snow. Buying hay means you care for cows, horses, sheep, goats, or other critters who savor that dehydrated summer fodder. With hay you also live with fences, water lines, land, and a decent pair of deerskin work gloves.

Hay changes how you spend your free time. People who buy hay might spend a Saturday riding their horses, then end it with a barbecue and campfire at home, inviting friends to come over for their evening entertainment. A paradise to some, but a prison to others, because living with animals keeps your life at home. You can't gallivant without serious planning beforehand. If I want to leave for a weekend, it means setting up kennel stays, farm sitters, stove tenders, live-in working houseguests . . . and, honestly, that doesn't happen. Even Christmas is spent here while my family is hundreds of miles away. I'm okay with this trade-off. A day on horseback and guitar music by the campfire is a lot more appealing to me than the crowds, concrete, and volume of suburbia.

That said, sometimes even this stubborn hay-flecked trail rider pines for a late-night stroll down a city block and Korean food at three A.M. You can't do that when you're living the life of hay.

What is this stuff that changes the geography of families and entertainment? To be blunt, it's just grass: tall grass, cut, dried, and left in fields until it is ready to bale, stack, and store for winter days when nothing is green but pine trees and holly bushes. Do not confuse hay, which is food for animals, with straw, which traditionally isn't grass at all but the stems of grains like wheat. I always explain to new initiates of the hay life that straw is dead, yellow, has zero nutritional value, and is useful only for compost, mulching gardens, and animal bedding. Hay, however, is alive, green, and sacred around these parts. Every bale is a little prayer for summer's return, pasture preserved, just like canned late-summer tomatoes as sauce. There's no difference between opening a quart jar of marinara and breaking that July bale into flakes. It is food set aside for the months when food doesn't grow.

As wonderful as hay is, it is also extremely expensive. An order of fifty bales (about a month's worth of feed for this farm) can cost well over two hundred dollars, as much as a cord of dry, split firewood or fifty gallons of heating fuel. I try to get as much of it as possible by bartering, trading goat kids or sheep for their monetary value in hay. It's the one thing I feel I cannot get enough of, and always have to keep purchasing because my barn is too small for an adequate winter's supply. Until I can afford to expand my fences or create new pasture, even my small number of hoof stock depends on this put-up food source. So until those days of bigger fences and better management come, I am always traveling around to one of my three hay suppliers, buying ten bales at a time and feeding out of my truck or the supply in my small red barn. My pickups grow more frequent as August flows into September and October. I want at least a month's worth stored up by snowfly, more if possible. It's a heartbeat of panic, hoping you get enough and get it put away come the frost. I am green with envy of my neighbors who have larger barns and wallets and can purchase a whole winter's worth in a weekend. It is wealth beyond comprehension to me and this dented Dodge pickup, dozen sheep, two goats, and a pair of ponies. Yet what I feed all season couldn't sustain the dairy on the other side of the mountain for one morning's feeding.

I think of this, the whole religion of hay, as I wait for Nelson to come out to talk. I lean on the gates and look at his happy herd of Angus heifers. It was here that I found myself feeling like a local farmer for the first time. I was watching the heifers, and without trying, without thinking about the questions I was asking, I dove into a conversation with Nelson about his herd. Now that I was a few years into the agricultural life I could tell their age and their breed, and I knew enough about cattle to see they were pregnant and well cared for. I asked, "How much longer until calving? Do you artificially inseminate them or prefer the 'traditional' method?" At this Nelson opened his mouth wide, showing the open spaces where several teeth were missing, and guffawed like a Muppet. "By Jeesum, I'm a *traditionalist* all right!" and without nudging my ribs made it clear he was a fan of all sorts of reproductive activities. And I didn't even blush. Nelson has always been a horrible flirt, which is just part of his charm, and I looked for the bull among the soft-eyed ladies in the paddock. As my eyes scanned, I thought about how many younger farmers around here wouldn't even consider keeping a bull, ensuring that there is continued business for the folks who sell the long straws of bull semen. Not Nelson. The land he was farming was a dairy for generations, and he grew up with a father who kept a bull, and he would too.

Nelson talked about his girls with pride in his voice, and bragged a little about his bull. He had reason to brag. The lot of them were gorgeous, and in the late afternoon's cloud cover and light rain it felt like a scene from an Andrew Wyeth painting. I looked over at the metal silo just to our right, a behemoth at least five stories tall, and asked him if it was holding silage, since I smelled that sweet, chewing-tobacco wetness in the air. He said that particular silo was empty, but he used to fill a silo much larger than that when he was milking 160 cows back in his prime. He pointed up the hill to another barn and told me about how when he was twelve he saw an ad in the paper for free literature and building plans from a silo company. He requested the literature, and a few weeks later a salesman came out to talk to him. He must have been a honey-tongued devil, that silo pusher, because he managed to sell an eleven-year-old the plans and materials, and little Nelson (with the help of his family, I'm sure) erected a ninety-foot-tall silo. "It was the largest in the county at the time," Nelson said, beaming at me, "and everyone said there wasn't enough feed in Washington County to fill it. But I always did."

I tried to picture the people and landscape of our county back then, in the fifties, when a new silo could be the talk of a town. Nelson has lived such a mythic life I can't blame him for never leaving his home. He still farms it, and the place probably looks just as it did back then, in his childhood, with rain-speckled cows, tractors, cement and wood walls, and silos being lashed by the rain before it reaches down under the metal sheeting. I look up at the huge structure, which is now known as the "little silo," and said a quiet prayer. Accountants and graphic designers may not be forced to deal with their own mortality on a daily basis, but farmers are. Be it a cow about to calve, a steer headed for slaughter, or a silo now barren as a winter field, we're only useful for a little while. Best to stand tall even when we're through with the work of it.

Behind us twenty bales were strapped down on my dented

Dodge pickup. I didn't worry about the rain on them, and I knew my horses and sheep would eat them before any rot could even consider setting in. So we just relaxed into a weekday afternoon.

We were friends, Nelson and I, talking about our animals in the gentle rain. Rain came down, and cow tails flicked mud, and there we were, leaning against the gate and talking about his plans for a new hay barn, fences, and a herd he hopes will reach a hundred head in the next few years. I kept a poker face the whole time, but I wanted to grin as wide as a Cheshire cat. Not because anything Nelson was saying was particularly funny, but because he meant every single word he said and had every conviction to make those plans happen. This is not a common trait. But hearing Nelson talk, I knew I wasn't hearing pipe dreams or prayers—I was hearing exactly what was going to happen. This is powerful stuff.

When I waved good-bye, Nelson proposed marriage for the fifty-sixth time, and I smiled and headed down the road with my dog riding shotgun and a heavy load of good hay in the rear. My hair was wet under my wool cap and my sweater sleeves were damp as well. My chest was warm to the core under a canvas vest, and I blew on my hands a little to warm them up for the half-hour ride south to my farm. I turned on the local country station, sang along to a song I knew all the lyrics to, and was home and unloaded before many companies would have even let their employees off for the day.

Nelson is almost eighty years old. He is in the hospital on a regular basis. He has mornings he can't stand up, or breathe, and he lives alone. But this man is working hard, and loves his life. He is making plans most people in their thirties don't dare to dream of without doubting his own ability. His life is something that keeps happening *because* of him, not *to* him. I love this about him. I really, really want to be like him when I grow up.

RIOT

COLD ANTLER IS WELL OVER SIX ACRES, but nearly all of the action happens within a hundred feet of my house. The feeding stations for horses, sheep, goats, chickens, rabbits, and pigs are all within a few paces of each other and my home. This may sound crowded and smelly, but it isn't. Think of my home as the center of a pie chart with all the animals' homes and fences spreading wider as they get away from human quarters. So the critters can walk up to me from their fenced paddocks and fields to pick up dinner, but their recreation and personal barn space is farther afield. This makes chores a lot easier for me, too, since the distance I need to travel with buckets and bales is short. This proximity also has its downsides. They are most notable around five-thirty in the morning.

See, that same distance that makes feeding horses, sheep, goats, chickens, and pigs such light work also means they are close enough to the house to know me as well as I know them. Those of us who raise animals often forget that animals are paying attention, learning our habits, and keeping track of everything in their sphere of activity. I have learned a few ways to outsmart them, but not many, and not without major inconvenience to my own routine.

For example, the horses know when I wake up. As soon as lights flicker on in the downstairs kitchen I can hear the heavy bellow of Merlin, who knows that since I'm awake I am therefore perfectly capable of delivering his breakfast rations. The sheep have also caught on to his trick, so when Merlin starts his demands, they join in. This is not a lovely sound. It is nothing like the childhood farm animal toys we had that let out happy rooster crows, whinnies, and lamb bleats when we turned a knob or pushed a button. This, my friends, is the preamble to a riot, and only a few strands of electric fence stand between me and a hungry mob.

There's nothing bucolic about this wakeup call—large, hungry mammals have my number. So if I want to start my day without a thunder of demand I need to wake up and head downstairs in the dark. I can't light a candle, slam a door, or light a match. If I do you can bet a horse will raise its clever ears and start the morning chorus. Every now and again I outsmart them, but it's nearly always in winter when I do. In the summer the open farmhouse windows let out every cough and conversation and I admit defeat without even trying to quiet the masses. So before the cats are fed or the dogs are walked, I head outside in the dim light of predawn to carry a few flakes to the horses to stave off complaints from the neighbors. The farm across the street and another up the mountain both must be able to hear this ruckus, and to their amazing credit they haven't walked over to give me a talking to. Yet.

HOWLERS

COME MAY, THE CHICKENS I had carefully raised from innocent, Easter-greeting-card fluff balls have left their adorable phase. There is nothing attractive about adolescent chickens. They have feathers and height, but look like tiny dinosaurs in feather boas. They now boast long, scaly legs and too-large beady eyes, and stalk bugs and get into boxing matches with each other along the hedgerows. A few weeks out on grass in the movable hutches or open-bottomed "tractors" (if they are meat birds) or ranging free around the barnyard (if they will be laying hens) has turned them into awkward, miscreant youths. They lack all the happy, matronly roundness of mature hens, as well as their calmness and industry. Instead of sauntering through the fields, cooing between pecks at bugs, these little hooligans are teenagers on the move. They do not walk, but instead they *run everywhere.* If there's a chance to make a noise, they make it, and loudly. First-time crowers lift their heads to the sky like wolves and let out moans only a mother could love. They sunbathe, but only in short bursts in piles of dry earth, where they stretch their fast-growing wings only long enough to catch the shortest acceptable amount of solar love before erupting into epileptic dust baths. They boldly jump into the pigs' pen to steal from porcine dinner plates. They jump on the backs of sheep and goats.

It is madness.

With their new plumage in bright colors combined with their antics they look like an eighties punk band or an anarchist's collective that has taken over my otherwise bucolic setting with a devotion to their own idealism. In this case, that idealism is nothing but spent energy and attempts—poor, poor attempts—at sexual congress. The immature males climb on top of the females after displays of bad dancing and horrible crows and make a few stabbing attempts at the end game but are usually sideways, or too slow, or just embarrassingly inexperienced, and the young pullets lose interest and walk away toward the stream or to scratch an abstract design into the gravel driveway.

The more I think about chickens, the more it sounds like art school.

Despite all the awkwardness, I can't imagine a May afternoon without this show. Chickens will always have a special corner of my heart as their permanent nesting box. Gardens are wonderful, and rabbits and bees have their charms and roles in even the smallest backyard homestead, but chickens make a farm. Ever present, their coos and warbling are a beloved white noise to the louder interjections of larger animals. Since all the other creatures remain contained by their pens, fences, and cages, the chickens seem extra gregarious. Their missions and antics around the raised beds, sheep fields, and horse paddocks remind me of quality-control teams or middle management, checking on the status of the other critters' work during the day.

When I walk outside for my afternoon check on the livestock, I usually just bounce between different groupings of chickens. I run across one rooster perched on the sheep's gate while a trio of his hens enjoys sweet feed dropped and fresh water from the sheep's trough. If they are calm and happy, and I can see the sheep are all accounted for, have water and plenty of grass and are minding the fences, I can move on to the next group of chickens, by the horses. There another rooster (there are always three to five of them here on staff) shows a group of mixed-age hens how to scratch open the three-day-old piles of horse apples that are filled with delicious maggots and flies. Between gorging themselves on bugs, they hop up on the electric fence—*zap!*—and I know it's working. I check the horses' water, give their body language and fences a once-over, and move on to the barn. Inside the barn are the two sleeping pigs and the goats, who are lazily munching on some hay. Chickens are hidden all over the nooks and crannies of the hundred-year-old building. They are laying eggs under the wooden handmade milking stanchion, on a shelf in the corner, or by the rabbit cages. One is helping herself to guest quarters, depositing her brown egg inside a bucket hanging over the goat pen wall. She reminds me of a mountain climber making camp for the night on a cliff face, photographed for a mountaineering catalogue.

By the time I have finished my rounds I have covered the entire farm and all the working animals that share it with me, and the chickens are the one constant. There are only twenty or so, spread out over the six and a half acres, but it seems like more, since they rarely stay in one place for long. They are not loud, and they are not smelly. They are fixtures as normal as the flies buzzing past me and the crows landing in branches overhead. Unlike the flies and crows, I offer them food and shelter in exchange for stealing their eggs (and the occasional chicken dinner) on a regular basis. So far the complaints have been minimal. And the saurian preteens running like a pack of velociraptors through the nettle patch chasing a luna moth keep reminding me that even the most serious and majestic rooster or responsible mother hen was once an idiot. Gives a gal hope that we get better as we get older.

ROGUE

SOME PEOPLE MARK the beginning of spring by the return of migratory songbirds, the first melt in the ice along the river, or those defiant little crocuses that punch out of the earth like anarchists' fists. My spring comes a little later; I mark it by the start of the annual sheep-escape season.

I can't really blame the sheep. The grass and brush growing outside their fences have been taunting them for weeks. The sheep have been sticking their noses under and through the woven wire fences to get whatever they can steal, and they know only more of the good stuff is just beyond their paddock walls. It's comical, really, the line they've traced around the fence. It looks like deer or very agoraphobic hikers had blazed a trail around the sheep's property line, but it is really all the work of stretched ovine necks who are sick of last summer's hay and aching for something green between their jaws. The flock has a large pasture, but I have split it into sections so the grass can have periods of rest, a time to grow tall and strong without ruminants tearing all of it to pieces.

So far I have been lucky. Only a trio of clever sheep has figured out that their horns don't feel electric shocks and if they can lift up the lower wire with their headsets, and quickly squeeze underneath it, their woolly backs (just weeks away from shearing) will protect their hides from any shock. To me this is genius work. So many farmers and layfolk go out of their way to call sheep stupid, but I have never seen any evidence of that. My flock knows me, my fences, their paddock, and their playbook. It was spring, and the game was on.

I'd finished the evening chores about an hour before, and it was just before dark when I heard the sounds of sheep calling. I've lived alongside sheep for years now and I have acquired an ear for what all the different bleats, baas, and cooing mean. This was an all-out holler, translated roughly into English: "Heeeeeeeeey, heeeeey! Hey, guys outside the fence! We want to be outside the fence with you tooooo. Heeeeeyyyyyy!"

The sigh I released must have weighed six pounds. I was inside; having just finished washing up the last of the dishes and swept the living-room floor. That last action is one of consequence. Sweeping the floor is the last chore of the night, something done more out of ritual than necessity. I have a beautiful handmade broom, which was a gift from my good college friend Raven Pray. She brought it up on a train all the way from Maryland one October, and every time I use it I think of her, and prepare myself to call it a night. Sweeping the last of the dirt, dog hair, and grass clippings I dragged in on folded pant legs is the chore that says, "Okay, kid, you did good. You can sit down now." And that is exactly what I was doing when I heard the sheep call. I was sitting in an overstuffed chair by a roaring woodstove curbing the late-spring chill in the night air. I had poured a Guinness into a tall glass, my DVD player ready for a movie, and had changed out of smelly farm clothes and into house clothes. All of this must have been teasing fate because those ovine hollers outside meant some sheep had escaped. The rogues were outside the fence and were being called to by those inside. With my luck the refugees were in the public road by this point, a dangerous position in the evening on a winding country road. There was no question about what had to happen next. I had to get dressed and head back outside. Good-bye, fire. Shut up, mocking broom.

It was colder out, around forty degrees with wind, and the weather report was calling for frost. This had me frustrated. May was not a time of year for erratic weather; it seemed an act of defiance above its pay grade. Late-season frost was April's job. I was back into a sweater and worn jeans. My feet were in still-damp wool socks not yet dry from the evening chores. You can point to a line of horse stalls and hand me a shovel or a dozen gardens in need of hand weeding and I will do so without a bristle of complaint. But ask of me the same amount of effort in wet socks and you are left with a sorry-looking animal.

Evening chores had taken longer than usual that night. The combination of cold-weather and warm-weather chores became an emotional thunderstorm. A farmer needs to be quick on her feet, fitting in the usual spring milking, weeding, chick wrangling, and bottle-feeding of kids on top of the work of chopping firewood, protecting plants from coming weather, and starting and maintaining a fire. After the animals were fed I had the extra work of watering and then covering with old sheets all the garden beds that I wanted full clemency for. I could now see the rogue sheep behind the fences, up in the woods and along the thick bushes and brush by the roadside. I stood outside my house by the lamp post and called to the sheep. "Come here, you wooly bags of dim suet!" I yelled, borrowing the insult from a favorite book. And then the parade headed toward me. There may be frost in the air, but as far as the sheep are concerned it's time for spring.

Maude led. She's the oldest of my ewes, a purebred Border Leicester of stunning white. In her full wool coat she bounced down the hill, the crescent moon hung in the sky above her. Behind her in a perfect V came six other escapees. They all trotted with heads high, horns gleaming in the lamplight, and fluffy coats bouncing in step. I would have been angry with them if they weren't so damn beautiful. I heard a young animal's cry, and fifty feet behind them the youngest ram lamb of the flock came running from behind to catch up. None of the other sheep turned around to watch his fuss. They were like trained soldiers reporting back for duty. Maude stopped her troops a few paces ahead of me and looked me up and down. I was not holding up my part of the deal. It is understood between my sheep and me that if I am going to call them away from a dinner of lush green things I'd best have something even tastier on hand. I am not above bribing sheep, and they are not above accepting bribes. Maude, having seen no evidence of grain, started to turn away toward the chicken coop, where a feeder of chicken scratch was always available to the birds. I put my hands on my hips and called to her, staring daggers. She stopped and looked and then looked away. Before she could rally the flock into a more severe act of anarchy I walked to the back of my pickup truck and shook a bag of chick starter feed.

"Can we please stay inside until daybreak? Please." And I shook the bag toward their faces, letting out my best impression of a confident sheep matriarch. I baaed at Maude and her ears lit up, probably more in confusion than comprehension. She took some steps toward me, now that I had the goods in my hand, and the others followed her. When the sheep inside the fence saw I was setting out a buffet they ran down the hill to meet me. Now in the full dark of night I had two merging parties of sheep about to clash into each other like warring throngs. Grain, any grain, is a serious treat around these parts, and they were all willing to bodycheck and push their way to the prime serving position, which was right at my feet. I opened the gate and stepped inside. They circled me like sharks.

I led the outsiders back into the main gates and dumped some chicken feed on the ground, making good on my promise. One by one the sheep came back inside the fence to join Sal and the Cotswolds who hadn't escaped and were rewarded by getting first dibs on the feed. They ate with a determination I could not exaggerate; they looked like the multicolored plastic hippos eating white marbles in a board game from my childhood. All of this fuss was taking place by the horses' paddock, and the two equines had trotted over to us from their pole barn. Merlin and Jasper watched the parade behind me, heckling me themselves. Jasper seemed to give up and trot back to the barn first, accepting that no grain was coming his way. But Merlin stood fast, stomped a hoof, and stared me down the way I had stared down Maude. He reminded me of a toddler demanding cereal in a grocery store aisle.

I have no idea what really goes on inside a horse's head. If he understood the series of events that had just occurred at their most basic level, he saw me rewarding acts of anarchy with abundance and pleasure. He had not escaped—where was his reward? He huffed and I patted his nose and told him, before walking toward the sheep's escape hatch in the far fence, "None of this is fair, Merlin. But it's ours." This pseudo-aphorism meant nothing to the beast, and he stared at the sheep's enthusiastic evening snack from behind his electric fence.

Before I could head back inside to my warm beer I walked up the hillside to where I had guessed the sheep had escaped. I soon discovered the bit of ripped-up fencing. All the sheep have learned that with enough force (and a low enough electric charge) they can shimmy under a weak piece of fencing once their head and shoulders squeeze through. Others in the flock see this act of rebellion and imitate it as soon as they have the opportunity. This particular escape hatch looked like something out of a cartoon. A perfect half circle of broken wire leading to a trail through the tall grass to the road I had found them on.

Without the aid of a flashlight I ended that evening's second round of chores repairing holes and hatches in the poor-quality fencing. I did this with my pocket knife, baling twine, and a few loose boards that had been propped against the sheep's small shelter barn. The last thing I needed was a school-bus driver beating on my door at six A.M. to tell me to move my livestock out of the road. When it was good enough I turned back down the hill, now sporting a light sweat and finally realizing how long the day had actually been. But I wasn't thinking about the movie on the coffee table, I was thinking about the weak electric fence and how to recharge it. What I needed was a clean, fresh string of electric wire right at sheep-nose level. I had a new grounding rod on order at the hardware store and plans to do it the coming week. But for now, for tonight, twine and a prayer would have to do. It's still all about action and reaction around here, not ideals, even in the best of times.

SHEARING DAY

I WAS RAISED IN A CATHOLIC FAMILY. Every single Sunday at ten A.M. my parents and my two siblings would go to Mass. We would sit in the same pew, third row from the front, at stage right, without fail, looking good doing it. While I hated dressing up in the little bobby socks and wearing tights and dresses, I must admit I looked adorable. My hair freshly crimped, held back in a plastic clip, I was almost as pretty as my older, blonde sister, Katie. I sat between her and my younger brother, John, on the wooden pew, my patent-leather Mary Jane–clad feet not touching the ground, looking through the only book in front of me, the hymnal.

My favorite hymn was called "Here I Am, Lord," and I think (though I'm not certain) it was printed somewhere toward the back pages of the book, in the early eight hundreds. Every time I read that hymn I felt empowered. It was a romantic ballad. It told the story of a man talking directly to God, and I thought that was neat. Talking directly to God was uncommon in the church itself, which in my memory was conducted more like a book club with just one speaker. There would be a reading, followed by an interpretation, followed by some rituals and shared communion, another song, and it was time to head home for *Crossfire* blaring on the TV while my dad made us all toast, bacon, and eggs. But that song wasn't anything like the rest of the church I had come to know through catechism classes.

The song was more like the homespun version of Catholicism my mom raised me with, which was perfectly okay with direct contact. My mom was always lighting candles and saying prayers, and she kept crucifixes and rosaries in everyone's bedrooms. Holy water was always on hand. She would dip plastic bottles into the wall fountains of holy water at the church and keep it around the house. Whenever I moved into a new apartment, she'd spray it around my new space to bless it. In other words, in my family, church members were allowed to talk directly to the deity and use magic. Which was a lot cooler than I realized at the time.

I was active in the church as a child. I was the first female "altar server" in the congregation's history. In the Catholic Mass, children (traditionally boys) get to wear robes and help carry sacred chalices and incense to help in the priest's rituals. When it was decided in the 1990s that girls were also capable of holding sacred objects in public, I was allowed to take the classes to become a part of the Mass. It wasn't a big role in the ceremony, but it felt like I was a part of something special. I was up in front of all those people in my community, helping with something I knew people felt was holy and important. I guess I should have known I wasn't cut out for Christianity if I was more excited about being in what I thought of as a performance than about taking part in a sacred rite, but at the unsavvy age of eleven I just thought it was neat to be a girl doing something only boys used to do.

Every year at sheep shearing I feel like that altar server again. The sheep shearer Jim McRae will be arriving to perform the annual task, and I, not yet having the knowledge nor inclination to do it myself, will be there to assist him. Once again I am a passive participant, watching a man who has dedicated his life to his work overseeing a flock that probably wishes it were somewhere else.

When I started raising sheep I had only three, outside my rented cabin in Vermont. They arrived in the late summer with short coats and throughout the winter gained thick inches of wool. As spring turned warm fast, I watched the wooly creatures hide from the sun in the shade of the pine trees or the small wooden eight-by-four-foot sheep shed I had provided for them. It was clear they needed a haircut, and as a complete novice I realized I had not lined up a shearer. So I did what I always do in times of farm need: check Craigslist.

I asked the Farm/Garden community for someone in the area who was a quality sheep shearer and who traveled. Not just one but several strangers sent me Jim McRae's name and number along with high praise. I called the Vermonter, who lived near Rutland, which was about an hour north of my rented cabin in Sandgate, and left a message. I wasn't sure whether anyone would travel an hour to give three sheep a buzz cut or what it would cost. I explained my need and prayed he'd call me back.

He did. And he explained that with small flocks like mine he would call me back when he had enough interest within the area. Coming all the way to my farm just to shear three sheep (at $6.50 a piece with a $25 flat farm-visit fee) wouldn't even cover the gas. So as soon as enough local flocks filled his dance card he would come to the farm and happily shear the two wethers and my surly ewe, Maude.

Having my own sheep shorn that first year felt beyond special. I had spent years reading about sheep, and finally to be involved in that timeless agrarian ritual—even as a bystander—was so emotionally overwhelming that I nearly teared up in front of Jim. I was certain that wrestling with fat Sal in his lap was not as endearing to him as it was to me. He sees thousands of sheep a year. I wiped my eyes when he wasn't looking and made up something about having a possible lanolin allergy.

I had collected sheep books, contacted sheepdog trainers and breeders, attended workshops and classes, and owned issues of *sheep!* magazine before I ever had reason to call a sheep shearer. I imagine it's how people who've dreamed of horses since childhood feel when they finally call a farrier to shoe their first horse, something utilitarian but wonderfully specific.

I watched Jim do the work of shearing and helped where I could. In this small a space with so few animals there wasn't much to do beyond picking up the fleeces and carrying supplies. He was the one doing the holy work, and once again I was busy as an altar server. Only this time I felt like the people in the pews, too. All of it was sacred to me.

Jim certainly is adept with his shears. It takes him less than ten minutes to grab a sheep, flip it on its rump, shear the belly, sides, back, and head in a boot-camp-ready buzz cut and trim their hooves. Once I got more experience, and every year after that first shearing, I helped more: grabbing animals in the holding pen and taking them over to Jim's shearing platform outside the scrappy fences. Once he has them it isn't long before the animals are shaved and set free of our clutches to commiserate with their flock mates on the hillside.

Seeing them under the apple trees, their newly pedicured feet in the mud and moss, I have to remind myself they are still the same animals. They look so foreign after so many months covered in wool, I forget that those black-and-white-speckled deer on the hillside are the same frumps I knew that morning.

I've been on a path of going from being a beginner at something—in this case, raising and breeding sheep—to actually doing it, making it a part of my regular life. It's been quite the adventure getting here. I went from being a gung-ho novice sheepdog trialer with an anglophilia crush on British breeders to someone making the mortgage on her own farm. This place has steadily grown since those first three sheep in a pen outside a rented cabin. The fifth year Jim came for shearing, my flock had grown to eleven sheep and I had the most wool ever ready for the fiber-processing mill. There have been mistakes, animals that died. But over fifteen sheep have also been added to the world because of my work here, all of them used to better the farm and my life through barter and swapping. Now there are Cold Antler Farm blackfaces as far north as Lake Placid and as far south as my friends at Common Sense Farm. The breed lives on in a world bigger than my backyard. Sometimes I forget that.

I don't think I'll be heading back to Mass anytime soon. I appreciate those memories, but I am best leaving them saturated in nostalgia. I'm not a Christian and certainly I am not cut out to serve any more priests, but I do value the time I spent there, having something so specific to compare to my current life of wearing lanolin-soaked pants and trading ram lambs for woodworking projects. Honestly, I don't see a difference between that form of worship and my own. In both, the worshippers are earnest people with deep gratitude for what has been offered to enrich their lives. Catholics attend Mass to celebrate and learn from a given sacrifice. I farm to celebrate and learn from the humble sacrifices made for six and a half acres of soil, blood, birth, and sex. The details seem inconsequential.

BLOSSOMING

AROUND SHEARING TIME is when I notice the apple blossoms. They appear slowly, each coming into the world as an individual, but somehow they always surprise me as a group at the same time. On the hill where the sheep live is an old orchard, a stand of no more than half a dozen ancient trees that still produce fruit. Some of the trees have been killed, the bark destroyed by an old goat I had named Finn. Those skeletal trees destined for firewood still stand among the living, and when the green trees burst forth with thousands of white blossoms they are adorned in their entropy by a gentle rain of white petals. It is stunning to see this, how an errant breeze picks up a few branches of the white fluff and sends it swirling into the air. It's magical. It's haunting. It's a haiku in the wind. . . .

And all I can think about is booze.

When I was brand-new to living this far north, some friends invited me to a pressing party. I wasn't sure what they meant by this. Did they want to have a really important discussion of current events? Press what? "Pressing party" was not a gerund-noun combination I had ever heard before. They might as well have invited me to a "lifting dance" or a "shoving potluck." I asked what a pressing party was and learned that "pressing" was meant literally, as in pressing apples to make cider. These new acquaintances were inviting me into a club that meets just a few times a year, in late autumn when the roadsides are bursting with wild fruit and trees in orchards are groaning. This is when these acquaintances assemble their buckets, baskets, garbage bins, and pickup trucks and drive around the countryside, from fence-lined back roads to established orchards, asking to pick up the scraps. "Drops" is what they call them in the business. We'd get permission from some people, others not so much, but we sure collected a lot of fruit. This mish mash of quality, from the most magazine-cover perfect orchard apple to the puckered wild fruit picked up off a roadside, made one amazing recipe for hard cider. It was with these scavenging apple jackals that I learned the song and dance of autumn hooch making. Using a friend's restored cider press dating from the 1860s, we crushed, pressed, and turned our bounty into twenty gallons of fresh cider. Add a little honey, some champagne yeast, and set it aside in an inexpensive fermentation vat, and in a few months that juice becomes one potent libation. Brewed with friends in the splendor of a New England autumn, it changed forever how I see those white petals. They are pretty, they are truly an auspicious sign. They remind me of friends, good times ahead, and horrible headaches that might follow nights without a hint of regret. Campfires, bonfires, Hallowmas potlucks, and harvest dances. The white blossoms make me think of all the celebration that comes along the apple road.

As these blossoms turn into fruit, a whole summer of life happens on the ground below them. Lambs grow and learn to be sheep, chicks follow fat mother hens looking for protein-rich maggots in the horse and sheep dung. Bluebirds nest, crows perch, and somewhere down the hill I am in my meditation of chores, writing, and worrying. The apple trees see it all.

And they shed their white flowers at the same time my sheep offer their own white coats. I live in one small place, on just a handful of acres, but the longer I stay the larger my world becomes. These trees are part of a story. A story that started with the farmer who planted them, never realizing I would come along, never knowing the future in their boughs. And a year of meat, milk, yarn, vegetables, and eggs, and the drama of a simple human life is happening below them, of which they are now a part. The blossoms mean that if I can keep the mortgage paid, keep the place running, keep my head on my shoulders, I will be rewarded before the first true snowfall. Winter seems so far away under a spring canopy of petals, but it is as certain as gravity, and there's a lot of work, bills, and sweat between the two white ground covers. It's as hopeful as it is terrifying. All I know to do is keep farming, and so I do. My part of the bargain is to tend and fuss, the apple trees' is to quietly grow and thrive. They do what they do and I do what I do and perhaps in the fall we'll both cast our shadows in the light of a Hallowmas bonfire and know we made it through another year. A circle is a fine religion. It keeps me going.

TOUR OF THE BATTENKILL,
FARM-GIRL STYLE

IF YOU WANT A LESSON in living locally, ask a working horse to teach you.

Every spring there's a big event here in Washington County, the Tour of the Battenkill bike race. Bicyclists from all over the country convene here with their army of fans and followers to participate in a 120-mile course through our farmland. It's both an amateur and professional endeavor, with weekend warriors and famous cyclists alike. For a few days the sleepy town of Cambridge becomes a festival, and thousands of people in very tight clothing driving extremely clean hatchbacks take over. I'm not a cyclist, motor or otherwise, and so I generally cut the town a wide berth that weekend. I can avoid the hubbub of town, but not the race. Three thousand people weaving through a hundred back roads in every direction is hard to avoid unless you want to spend the entire weekend in your living room. Even then, there's a good chance you'll see a pack speed by your property.

Not that cyclists should be avoided, mind you. I may not be a fan but I do respect their verve. It's not every human being who wakes up and says, "Hey, today I am going to travel over a hundred miles across the landscape without an engine of any sort!" As someone who is beside herself with satisfaction if she manages a three-mile jog, the amount of physical dedication that must take floors me. Just because their Spandex freaks me out doesn't mean they aren't the next step in human evolution. Any fool can see that. I mean, have you ever met a 60- to 120-mile-a-day cyclist? They are as tawny and lithe as Amish preteens, calm as monks, pleasant to chat with, smart as whips, and they can drink like fish. Laugh at their tiny shorts all you want, they are winning the gene pool game, hands down.

It was on one of these celebrated weekends that my friend Patty Wesner had an idea. She thought it would be fun to take our horses and carts out among the race on an eight-mile round-trip to Salem, to get ice cream at the beloved Battenkill Creamery. We could watch the cyclists go by and wave from the buckboards as they pedaled madly toward the finish line. It would be exciting to show off our fancy horses in harness, too. And what city slicker wouldn't love the country charm of watching their spouse pump past a horse and cart on their weekend rural race? So it was settled, we were going to harness up and step out for the big race. We rallied some friends, polished the tack, groomed the horses, and prepared to hit the road grinning.

Patty's Percheron, Steele, was in his new harness, hooked up to a beautiful green and yellow wagon that comfortably fit four people. At nearly a ton, Steele was large enough to pull them without much effort. Merlin, my draft pony, was half Steele's weight but built just as sturdy. He was pulling a metal forecart, a lighter vehicle for two, and I was in the driver's seat. Beside me was my friend Tom, a farmer in his own right from Massachusetts. He was visiting for the weekend and had never been out on public roads in a horse-drawn vehicle before. If he was nervous he didn't show it, and I took that as a silent nod of confidence in both my horse and me. When all the harnesses had had their last-minute adjustments and everyone aboard had had their last-minute bathroom breaks, we flicked the reins and asked our beasts to take us away.

What with fall hayrides, paid carriage rides, special events, and theme-park attractions, many people have at some point been passengers in a horse-drawn vehicle. But riding along in a domesticated and slow-moving situation like a paid ride about town is nothing like being hitched up behind your own animal with friends in your carriage and heading out on an adventure. It feels completely different—almost subversive in how basic this form of transport is. It is rare, at least at this point in history, for people to take a horse cart instead of a car. If you're not living in the heart of Amish country it seems either archaic or charming, depending on your disposition. But for the lot of us it was a peaceful kind of thrilling.

We saw no cyclists for the first two miles, which were spent along the paved and dirt roads that connect private, hidden Lake McDougal to Black Creek Road. Steele and his wagon led our small parade, and Merlin was eagerly trotting behind him. I ride Merlin all the time—I had bought him so that I would have those adventures in the saddle—but on days like this I realize I have a driving horse that only tolerates me on horseback. And that isn't to say Merlin isn't a fine trail mount, but Brigit's Fire, does he move in harness! Without complaint or hindrance he kept close behind the wagon bearing our friends. Tom was having a good time, taking in the view from the seat of the cart and helping by flagging nervous automobile drivers past us.

People in cars and trucks do one of two things when they see a horse-drawn vehicle on the road ahead of them. They are either overly cautious or wicked. Some slowly rumble behind us, thinking it is dangerous for all concerned to pass on the left. And the other kind of driver just roars past without any consideration at all. They must assume that if a horse is on a road it must be treated like any other car. This isn't necessarily untrue, but if you ever see a horse cart in your path, treat it like a car, but a car driven by a little old lady. You can pass her, but do so politely. Get the job done kindly and efficiently by cutting her a wide berth. Don't speed past with inches of space between you, slam on your horn, or drive alongside her to take her picture.

Driving a horse cart isn't hard, but it demands a presence and instant reaction to changes. Unlike the bikes, cars, and motorcycles on the road, our engines have minds of their own, and that is a whole new layer of complication.

The sun was hiding behind clouds, but we weren't cold. A mild wind kept the horses' noses to the air, probably smelling the new foals and their mothers and red-tailed hawk nests high in the trees. The people on this trip were all chatting away. I sang occasionally, because that's what I am used to doing when out with Merlin. I only know the first two verses of "Loch Lomond," but I sang them with a horrible Scottish accent, and Tom and I laughed between the *clip-clops* of the two draft horses. The trees around us were still barren and only the eager poplars and some sun-luck maples were sporting buds that promised leaves. The stubborn locusts and oaks remained as bare-branched as if it were the dead of winter, and if it hadn't been for the occasional puddles and mud under the horses' strides I might have believed it was an oddly warm day in January.

However, all comparisons to winter faded at the first pack of cyclists. They were pumping up a hill from behind as we turned onto Black Creek Road. We weren't sure how the horses would take this. I wasn't concerned, but I also wasn't a one-ton herbivore strapped to a metal contraption pulling weight on paved roads. In a flash of neon Spandex and the metallic fizzing of relatively tiny, fierce wheels the mob whizzed past us. Merlin didn't even lift his head. I don't know if Steele was as calm, but he didn't bolt and the wagon ahead didn't stop, so I assume he was equally unimpressed. This was good because in the next mile or so that it took to get to the creamery, we would be passed by hundreds of competitors in the race.

You know that old Chevy Chase movie *Funny Farm*? The one where a couple from the city buys a little place in Vermont and it turns out to be a disaster? They eventually want to sell the house, but to do so first they need to pay everyone in town to act like Norman Rockwell characters so that any potential buyers are convinced they just walked into Mayberry. That's kind of what it felt like as we drove the horse past the checkpoints in the bike race. Tourists from out of town took our photos and squealed as if we were part of the show. I grinned and waved, but I felt like a local had paid me to put on the rural charm, when all our gang wanted was some wind in our faces and an ice cream cone.

We didn't know that the creamery was to be a checkpoint for the race. As we trotted the horses up the small rise to the entrance, the crowds thickened and more and more flocks of Spandexed pedal pushers started gaining on us. We asked the horses to move faster, and moved out of the way of the race by turning into the creamery's parking lot. Once the horses were set into two parking spots Patty offered to hold their heads while the rest of us went indoors for coffee, ice cream, and a bathroom break. While waiting in line for a scoop of coffee ice cream in a cone, I could look out the windows of the creamery and see strangers coming up to Patty to ask about the horses. Patty glowed like a proud mother at a talent show. She let kids pet the horses and fielded questions. By the time I walked out to stand by my own black steed I could hear the current conversation. Were we with the race or just locals out on a jaunt? When we explained we do this all the time, their expressions went placid with contentment, like they had somehow wandered into Narnia and had had any lingering doubts regarding their whereabouts dispelled.

We ate ice cream and watched the race. It was a good break for the drivers and the horses. Neither Steele nor Merlin seemed to have worked up much of a sweat, but we still had another stop to make, plus the return trip home. I sat on a picnic table with my prize. The ice cream was, as always, delicious. Battenkill Valley Creamery is a little gem. They sell milk right on the dairy site, in glass bottles with antique-style labels. You are free to walk out to the calf barns and see a half dozen little Holstein babies romp and play together. Some dairy farmers keep the calves in their own sheds and confined spaces, but not here. It's like a raucous summer camp. Tumbling and bleating abound while their mothers and several other generations watch from the hillside of stream, grass, and rocky ledges. To be so close to your food, to look it in the eye and see it compete for King of the Mountain as a brown-eyed babe, never loses its power. And somewhere midscoop, while licking the edges of the icy treat, I remember that this is not a special event for me, not a paid-for experience. (Well, the ice cream was a paid-for experience, but not the countryside trek via horse cart.) My life, thanks to a pile of good and bad decisions, had led me to a place where this is as normal as hopping into the car and going to the mall.

We got back into our rigs and said good-bye to our new friends. We waited for a lull in the race, between clusters of cyclists, to head back onto the road. When the coast was clear we trotted off and headed along a popular country highway, Route 30, to Gardenworks.

Gardenworks is an old Scottish farm set into some rolling hills. It specializes in You Pick berries, growing acres of raspberries, strawberries, and blueberries that folks come out and pay to pick. Fruit doesn't get fresher than when it is your own hands that have plucked it off the branch and brought it to your own front door. However, berries alone do not make the operation that is Gardenworks. The large barn adjacent to the berry fields—just twigs and rows in late April—is the site of an artisan market and gallery. The barn's main level is dedicated to local meats, cheeses, desserts, and preserves and local arts and crafts. The upper loft has whitewashed hanging drywall where art is displayed among old plows and threshing equipment. If you were an artist who did anything even mildly agricultural this would be the perfect place to display it. And to pull up aside this sunny barn in two horse carts felt correct and happy, something that was simply supposed to happen on a weekend as festive and community-centric as the big race weekend.

I stayed out with the horses while my fellow travelers went inside for a second coffee. Mark, Patty's husband, eyed the plants for sale outside. It may have been only April, but six-packs of lettuce greens and some bright flowers were already available for those eager enough to gamble with the weather. Late April still left a lot of time for frosts in these parts, but Gardenworks knows people around here are thrilled to take a chance on some lettuce after the long winter. If I had had any space on the forecart to pile in a flat of those blessed greens I would have taken them home that instant. *I bet they sell out of those six-packs by the time the race is over,* I thought. Just then, some tourists to the area for the race walked over to ask about the horses. We chatted and I leaned back a bit toward Merlin's head, giving him a scratch. As I told the visitor about the horses, the community, and the amazing Nuns of New Skete cheesecake for sale inside, I could feel Merlin's breath and smell his sweat, which had made his underbelly wet. He was breathing deep, working hard, and I took note of it. Breaks like this are good for both the driver and the horse because it gives us time to relax and catch our wits and breathe between the constant reaction and focus of being an animal-drawn vehicle on a road. But stops like this are also good for the people in this county, to chat and smile and share stories and buy coffee.

I am grateful for my pickup truck and all the work it does, but I don't have to stop and let it rest. I can move so fast, so concentrated, past businesses and neighbors and never stop to share a mug of coffee or tell a stranger about cheesecake. The pace that driving horses gives to your life is a reminder and a gift, one I am constantly grateful for. Having these animals in my life has helped me meet so many neighbors. On the little mountain road where my farm is located folks do not stop to talk to me if I am jogging, walking my dog, or driving my truck. But if I am on horseback or in a rig they always pull over and ask how the farm is doing, how I am. I think the horses make me seem more open and friendly, nostalgic and timeless. Folks see a rider or cart in the road and perhaps they are reminded of a different time and place, and part of them wants a taste of it, too. So they roll down windows and wave hello and ask about the new goslings they saw following the geese or the baby goats running with Gibson past the house. We talk with the comfort of old friends, even when we don't know each other's first names. The situation of a horse and a country road is enough to infuse us with comfort.

After our second rest of the day we got back into the rigs for the last time and asked the horses to trot us home. It is four miles to Patty's farm, and we take it slowly. We have sun on our faces, a bit of weariness in our voices, but it's all happy forms of wear. If people in cars are in cages, we are range animals. We accept the sunburn, the road dust, the pain in our rein-holding hands because we have learned a bit about what traveling really is. It's not just about a means of getting from A to B. It's not about showing off fancy horses for a local event. It's certainly not about making good time, saving gas money, or even exercising our horses. It's about the energy and people needed to see the world, even this incredibly local piece of it. We traveled an eight-mile loop, and it took half a day and several stops. We knew every part of our harnesses, our horses, and our rigs and had the means to make them all move us from one place to another. It involved time for conversation, song, laughter, and stories. We met neighbors and started conversations with strangers. A scoop of ice cream became road fuel and not a guilty splurge. That is a lot to gain from a pair of horse carts plodding down the road during a bike race.

BLACK LEATHER, SADDLE LEATHER

I HAVE NEVER DRIVEN A MOTORCYCLE. I've been on one, once. I'm not sure that time really counts, though, since I wasn't driving or even sitting behind the driver. I was sitting in a sidecar. I was offered a taxi back to the office from a coworker who owned the bike. The sun was shining, the early summer grass was speckled with wildflowers, and the idea of jaunting along in a little Russian sidecar seemed whimsical. I remember sliding down into the tiny metal egg and getting myself comfortable. It had a small seat and wood-paneled floor, which I thought was a classy touch. Then I asked the driver where the seatbelt was. Through a smile that was actually more of a grimace he explained that if you are in an accident in such an apparatus, the last thing you want to be is trapped in it. I believe the phrase he used was "thrown clean." That was what you wanted to be. A phrase you don't expect to hear on your lunch break. The ride turned out to be delightful and easy. And for a brief time afterward I toyed with the idea of getting a motorcycle of my own.

I have a theory that a large portion of motorcyclists are actually equestrians at heart. They strive for the same feelings of joy, speed, escape, meditation, and fellowship any band of trail riders on their quarter horses feel. But since horses seem to carry a burden of high expenses, land, and effort of care and feeding, many folks who would love to feel leather reins in their hands are drawn to motorcycles instead. And the more into their bikes they are, the more I feel it supports my theory. To ride a horse is to control great speed, power, and skill—so is riding a bike. The constant maintenance, oil changes, washing, waxing, coats of paint, meet-ups, riding events, clothing, gas, parts, and so on are probably just as expensive as any trail pony could ever be. And those folks who baby their bikes would lavish the same intense care and feeding on their steeds. No, it's really not so much about money as it is about commitment. The desire to be a cowboy is contradictory to the freedom of not having to return to the corral every night. A bike doesn't need to go home. It doesn't need to go anywhere. Which is exactly the appeal. And unlike a horse, it has no mind of its own with which to decide to bolt, buck, or simply get tired and stop. Motorcycles are still horses, but in a romantic sense. The people who ride them can feel that. It could be a blatant choice, or buried deep in their subconscious, but I stand by my observation that a lot of people in black leather really just yearn for saddle leather.

The only reason I considered owning a motorcyle was because owning a horse seemed impossible. The horse was my dream, and the dream was fueled by childhood books and movies. It didn't matter if the drug of choice was the horse Black Beauty, or the horse of Pippi Longstocking or the Scottish hero William Wallace—I wanted it. I can still see the mythic scenes of these characters running free in my head. And when you take a small-town girl and show her moving pictures of what is possible in the world, it leaves a mark on her. I watched those movies and those riders with wide eyes and a fast-beating heart. I ached for the transformative experience of moving across the landscape on a powerful animal I loved, trusted, and understood. It was a superpower. Who needed to fly when you could soar?

It doesn't seem that long ago that I contemplated becoming a biker back when I was leasing a little cabin in Vermont (before I bought my current home). I could barely afford three sheep, two dogs, and a flock of birds while still making the rent and car payment and keeping the place heated. Back then I assumed owning a horse not only took a lot of land (that you owned) but a lot of money. Horses were for rich people, or lucky people who had a cousin or a friend who let them board their horse at their farm. Horses were like being a batboy for the Yankees. You could physically do it, but the chances of actually making it happen meant you'd better know someone.

Since owning a horse was a pipe dream, I started looking into small bikes for shorter women. But since my passion really was for animals, and not machines, I didn't do much groundwork beyond some conversation, lackluster Craigslist searching, and picking up a motorcycle pamphlet at the DMV. For a while the dream of riding wild—on horse or engine—was set aside, and that pause in action was exactly what I needed for destiny to unfold. Over the next two years a whirlwind of luck and chance forced me to leave my rented land and led me to buy a new farm in a new state. I hadn't been on my land a year when I found a working cart pony for sale a few towns over and bought him. Jasper was a small POA pony that had been trained by the Amish for riding by children. As a young farmer with nostalgic passions for working draft animals I intended the pony to be my equine training wheels. The relatively small 450-pound animal could teach me about harnesses, equipment, horses, halters, and more while I worked up the chops to buy an animal I could trail-ride someday.

I had no idea what I was doing with Jasper. I knew enough to ask the right people for help, and found a good farrier, a riding instructor (to gain general horse knowledge), and some folks who excelled in driving. I asked my friends with horse experience about fencing, equipment, and other issues. I wasn't about to ride Jasper across an open prairie, but a small part of me ached for it—the allure of the saddle was so strong. To go from dreaming about riding a horse to finally owning one, I went out and bought a saddle. It was a roping saddle, size 15, and when I put it on Jasper's back he looked like a turtle in its shell. With resignation I realized that aside from the occasional pony ride for small children, my charge was going to remain a cart horse.

And a cart horse he was! Jasper turned out to know what to do in harness. He was green but willing, and together we pulled logs from the woods, rooted out honeysuckle and briars, and went on walks down my little winding country road. Just walking alongside him on those outings made me feel wealthy. He might as well have been a Thoroughbred just off the track with a wreath of roses around his neck for how I beamed. He wasn't Black Beauty but he was mine.

Jasper sated my equine hunger for a good while. I kept working with him, going on our walks and such, and I got serious about riding lessons. I took a private lesson a week from a local dressage barn just eleven miles from my farm, and there learned more about horses than any book or walk alongside a cart pony could teach me. The animals at the riding stables were tall, majestic, and so well trained that a slight change in my weight or a tug on the reins could get results racecar engineers would kill for. It took a few months in the arena, but one late-summer day my instructor, Hollie, asked me if I wanted to leave the lesson ring and head out on a trail ride instead. I didn't hesitate to accept the offer, and the coming moments would change the trajectory of my equine life forever.

Hollie didn't tell me to saddle up a warmblood or quarter horse. Instead of those tawny, tall creatures, she told me to get out the stout Haflinger mare. My heart raced. I had never ridden the horse, not even in a lesson, but had always wanted to. Haflingers are light draft horses, thirteen to fifteen hands tall. They are built like Belgians in stoutness and have long blond manes. When I got in the saddle it instantly felt more comfortable than the horses I usually rode. This animal was wide and strong, short and sturdy. I am built more like a hobbit than a gymnast, and I appreciated feeling like I had met my match. We rode out into the summer sunset and headed for the hills.

I rode that draft across stream, field, and forest path. Hollie and I walked, talked, trotted, and laughed along the way. It was the first time in my life I got to explore the world on horseback, and I fell in love with it. The animal below me was safe, willing, and calm and besides taking the occasional snack from tall grasses was a textbook trail pony. After months of riding lessons and trotting in circles, I knew this was what I wanted. It could have been the golden afternoon light, or the golden horse, but I was hooked in a way no motorcycle could ever compete with. When we came back to the barn to untack and groom the horses I made a promise to myself that someday I would do this on my own. I'd ride across the good earth on the back of my own draft pony. I had gotten a taste of what was possible in the world. I wouldn't be content until I could gulp it down again and again.

I kept taking riding lessons off and on as the weather and money permitted. I kept Jasper out in the field with the sheep, and he seemed content enough. I knew he should have an equine companion for company, but that seemed like too large a step to take. I knew if I were to get a horse it would not be another cart pony.

It would be an animal like the Haflinger that showed me the trail. And taking on that second horse would mean upgrading my farm's facilities, too. A larger horse meant a larger barn, a paddock just for the horses, fencing, tack, gear, and the guts to saddle up and ride away from the safety of arenas and instructors. I was going to need more than cash and some horseflesh. I was going to need a mentor.

It might have been fate, fortune, or an effort of will, but a teacher appeared. I had met Patty Wesner at a book signing a year earlier. She had read one of the essays and felt a connection with its author. After a little research online she discovered that the woman she had felt a kinship with lived just six miles away from her. When she introduced herself to me she told me she had a Percheron—a large breed originally bred in northern France—and he was trained to drive. If I wanted to I could come over for a cart ride and learn about working horses. And right on cue, I came across an ad online for my dream horse: a black Fell pony named Merlin.

Fast-forward to the present: both horses are in the paddock behind my house, and my onetime "office" has turned into a tack room. A desk, computer, printer, and a few books and cameras are still there, but all around me are saddles, blankets, grooming materials, bridles, and a pair of harnesses.

It could have just as easily been a helmet with a face shield, motor oil, and a poster of a dream bike instead of framed prints from horse shows and ribbons. I never did find a reason to buy a black leather jacket—but even so, I still feel like a badass with a fifty-pound leather harness draped over my shoulders.

FIRE OF BELTANE

ON THE LAST NIGHT OF APRIL, I light a campfire at the farm to celebrate Beltane Eve. It is a very humble affair, nothing like the way people celebrate elsewhere. I mean, over in Scotland the Beltane Fire Society puts on a show that would make Lady Gaga blush, with fire dances, a May queen, and naked people painted red and blue dancing and reenacting old rites and stories of Scotland's past. It looks like a hedonistic romp, or something from the set of *Game of Thrones,* but the performance is actually put on by the Scottish Studies department at the University of Edinburgh. Over three hundred dancers and collaborators donate their time, and tickets always sell out. They know how to party over there.

If Ostara is a start to a new cycle of life, Beltane (or May Day) is the party to celebrate all that creates life, and the beginning of the pastoral summer season. It is unfiltered romance and energy, passion and joy. The world is lush and I feel vibrant. But here at the farm I don't paint myself blue or run around naked. I just don't have the time. Legend has it that if you jump over the flames on the evening before Beltane you gain the luck and favor you need for the coming year. I jump over the backyard fire with as much intention as I can manage. Even if all it accomplishes is making me feel powerful for a moment, or convincing me that better times are possible, it is magic to me. To change a mood or offer hope from a ritual is the basis of prayer around here. I feel my sturdy frame fly and land with the grace of a jungle cat. Sometimes I feel indestructible.

The day after this fiery leap of faith, an opportunity for introspection, I celebrate Beltane at Cold Antler by throwing the first outdoor potluck of the year. It doesn't matter if people see it as a religious holiday or not—it's just a fine time to collectively celebrate the start of the gardening and farm season. I open up Cold Antler to the neighbors for a vibrant outdoor gathering. Early May is warm enough to want to spend the evening outside, even if you need to slip on a sweater come nightfall. Folks bring what their gardens or winter storage have to offer—green salads and roasted kale dishes, potato salads. Sometimes I roast a leg of lamb from last fall's meat harvests. It's a heck of a meal, all from farms and friends around the area.

One memorable year I devoted Beltane to working with both the horses. I trained Jasper in the cart with some friends from Common Sense Farm and took Merlin for a trail ride up the mountains at sunset. I sang to him the whole time. I made up a song about May, and Beltane. Anyone who could hear me would have laughed at the silliness, but I was happy on that black pony, even if I was a little scared. Every time I get on Merlin there's still the fear of a beginning rider, that hesitation and understanding that I am about to mount a thousand pounds of powerful muscle with a brain of its own. To temper this, when we start up the trail I sing. It relaxes me and so it relaxes Merlin. A horse can tell if you are tense, and assumes it's for good reason—there must be a predator, danger, something to be worried about. When I start into a few verses of a familiar tune we both breathe deeper and enjoy ourselves.

Every day I get better with horses. I want to be good at this life, have an understanding that connects me to all those who jumped over May fires centuries before me. To hunt, farm, sing, and ride on a horse has become the definition of a life well lived.

People have told me that I am an old soul, because of my love for Scotland, archery, horses, hawks, folklore, and agriculture. Or perhaps I was a Scot in a past life—a romantic notion, for certain. It sure would explain why a girl raised in Catholic suburbia would find herself wearing a kilt, riding on horseback on the eve of Beltane.

My mother always said that if you can inherit your grandfather's eyes, or your aunt's red hair, then why couldn't you inherit their memories and interests? What is there in our DNA (which we only understand at a basic level) that stops who we are mentally from carrying over into the next generation? She doesn't believe in reincarnation, but she does believe that when you experience déja vu it would be an accurate memory because it may be from someone related to you in the past. Perhaps my desire to play the fiddle, ride horses, and raise sheep is just part of my gypsy bloodstream. Maybe far back in the forest of my own family trees there was a Scot in the Highlands, a tinker in Ireland, or an exile to Prague that slipped some of him- or herself down through the bloodline to me. I like to think that's why I am who I am, and how I will get where I am going. I prefer to think we are who we are through human actions like sex, decision, and passion than random rebirths at a deity's whim. At least, that is what I was taught, and I'm sticking to it. My mom always said, "Science can only bring us closer to God." That it's the exact same thing. The more we learn about the universe and earth the more we can understand our source.

I spend my springtime holy day with horse sweat and fire jumping. Perhaps you do the same. Maybe you plant peas and savor the sunshine on the back of your neck. It doesn't matter, it's all the same. And that's the most sacred truth of all.

SUMMER

GROWTH

THE BREAKFAST CLUB

I WAS THE ONLY ONE AWAKE in the farmhouse. Or rather, I was the only human awake. Two hungry cats with me in the kitchen were ravaging their kibble bowls, and two dew-soaked dogs were romping in the living room, refreshed from their morning walk. Just behind the kitchen in my mudroom chirped two dozen month-old chicks in their comfortable brooder. So I was one of over twenty-five of the awake animals inside the farmhouse. And if that sounds crowded, I assure you it isn't. The little chicks are feathering out but are still small enough to be held in the palm of your hand. They were all inside on that late May morning because of a frost warning, and their outdoor pens might have been on the chilly side for any birds not lucky enough to be in the center of their slumber chick pile. So before bed I walked outside with a flashlight and scooped up chicks and carried them inside to a warm brooder with clean pine shavings and a heat lamp. Now as the morning sun was washing over the chilled grass I knew they were ready to get back outside. But before any chick would receive taxi service I was going to get my guests some decent breakfast.

The remaining trio of animals in the house was still asleep upstairs in the guest bedroom. My old college friend Sara had sent me an e-mail out of the blue a few weeks earlier, asking whether she and her husband, Tim, and their ten-month-old daughter, Juniper, could visit for a few days while on a road trip to New Hampshire. I had not seen her in years, and had only kept up with her through a few rushed e-mails and social-networking sites, but our sketchy contact had more to do with our busy lives than poor intentions. She was a new mother and wife to a first-year resident doctor in the Philadelphia suburbs. I was happy to let them sleep in, excited about it, even. I wanted them to wake up well rested and content, and when they descended the stairs I wanted them to smell percolating coffee, sizzling bacon, and scrambled eggs ready for them to savor.

I had turned on a little music, the volume low enough so only the downstairs crew would enjoy it. I was cracking eggs into a bowl and whisking them into a yellow froth with some fresh goat's milk. On the range a large pan of bacon smacked and danced with pops of fat. A pile of kale burst from the bag in which it had made the brave, three-mile trek from Common Sense Farm's fields, alongside head lettuce and leeks, its brethren. These vegetables were dear to us because all four of us humans (baby, too!) had been out in the field helping with a community harvest the night before. (When guests who are interested in farming visit my town, I give them the haycation they are looking for.) This morning's breakfast was a celebration of our spoils: payment in organic produce enough to feed us all until we were stuffed.

When we'd arrived at Common Sense the night before, each one was assigned a different task. Tim spent his time with other men in a distant field, cutting kale and piling it into large bins. Sara, Juniper, and I were assigned to the leek station. A few long rows of leeks were ready for picking. I used a shovel to loosen the soil and dig up the big leeks, pulling them out of the soft earth by their long stems, like a magician pulling a rabbit out of a hat. I set each down on the earth and then someone else came up the row behind me and shook the soil from the leek's roots. The unearthed leeks were set on a grassy pathway, and children and a few elderly guests gathered them to be put into large bins for washing and presentation. Once cleaned of all dirt the leeks had their long roots trimmed to a sharp buzz cut and their long leaves cut down into a pronged fantail. What had started as a literally dirty plant on a patch of grass now looked ready for a green market's photo shoot.

The time went by fast in those fields, as we told stories and caught up with friends. Everyone was as focused as they needed to be, but this was the kind of work that kept hands busy and stories leisurely.

Around us children of all ages were by our side. Far from being child labor, it was gardening, and everyone was having a grand time! Little ones less than a few months old and strapped into carriers watched their mothers in wonder as they sorted roots and rinsed off bits of soil. Toddlers ran barefoot between the rows, playing with kittens or discovering bugs. Preteens and teens weeded, carried boxes of clean vegetables, and ran messages around the five-acre plot. In the setting sun, the scene was instantly nostalgic, the kind of moment you might see in a movie about Utopian societies . . . but it was actually happening. These moments are rare around here, but when they do happen it makes the dirty parts worth the fuss. We came home from the farm tired and happy.

And looking at the perfect vegetables the next morning I knew my guests would know how special breakfast can be, and really is— leagues beyond cereal shaken out of a box. Sara, Tim, and Juniper were a part of this meal's story. They were on the team taking the food from ground to table. Alongside the gifted vegetables were eggs from my mixed flock of hens. The bacon was from my winter pigs. And even though I have grown used to eating food that I know this intimately, the fact that it keeps happening keeps amazing me. I grew up shopping at supermarkets and eating meat that came in Styrofoam packages. Bread was bright white and came in plastic bags, many printed with the name WONDER. Every time I think of that I laugh because back then it was just a brand name. Now, when I pull a loaf of fresh-baked bread out of my oven, wonder is exactly what I feel.

Sara came downstairs as I was transferring bacon from pan to plate. Shortly after, Tim and Juniper joined us. The babe was all smiles and clapping hands, lighting up the farmhouse the way only a baby could. I handed out mugs and poured dark coffee from my stovetop percolator. Sara asked what she could do to help, seeing as breakfast was already under way. I handed her a chef's knife and a cutting board and she started showing the kale and leeks who was boss. Soon I had the savory chopped leek rounds dancing in the bottom of a pan of oil, a big wok that was a hand-me-down from other friends. When the leeks were soft and browning a bit I added sea salt and the rest of the chopped kale. It filled the wok! With a little extra oil drizzled on top it quickly softened into sweltering pieces, glistening with oil, and the aroma of the fresh greens filled my nostrils. I cracked eggs into a big bowl and whisked in some milk. I poured the mixture over the green goodness and a messy-beautiful scramble became a melody. If some meals are symphonies, this breakfast was a folk song sang in rounds along a forest trail.

I scooped big portions into bowls and sprinkled a bit of grated cheese on top of each. All that with a crisp side of bacon or piled onto a sandwich bun made for a breakfast not soon forgotten. The sun was shining bright and from the kitchen window we could see the horses munching their breakfast. The twin newborn lambs were asleep, piled together on the hillside while their mother grazed nearby.

This is heaven, was all I could think, between bites of the blessed meal. This is heaven, and a sinner created it through effort and will. I took another bite and chewed it too long and too slow. I wanted to savor in ways people no longer savor. I had made a phone call to my bank the day before, setting up a mortgage payment. I was now the proud owner of the label 30 DAYS LATE on her mortgage instead of 60 DAYS LATE. It had been a tough winter and I had gotten behind on bills. As spring grew warmer a new frugality was my reality. I was crawling out of a financial hole as best I could, and although I wasn't happy to still be behind in repaying my debts, I was making progress. I was feeling grateful to be surrounded by beauty and a dream slowly coming true, even if I was broke.

I had told Sara the day before, while we were on a two-mile horse-cart ride with Merlin, that if I won the lottery or came into any amount of money I would work harder than I ever had in my life. I would pay off my debts, buy a larger plot of land, and teach people how to farm and raise their own food. I'd make music and bring old words and old songs back to life. I would shape my farm into a place where you could go back in time and be surrounded by people who wanted to be there with you. I would grow good food for anyone who wanted it for free. I could wrangle volunteers to take it to inner cities and food deserts where good veggies are harder to come by. I'd offer it to food banks, or anyone who wanted to drive up to the farm and take some. When money isn't being exchanged I bet it is easier to give the gift of meals like I was experiencing. I proclaimed this with conviction as the silver Celtic knotwork bell jingled on Merlin's harness. Living for yourself is good. Living for yourself and feeling free is great. Living for others is better.

This is the dream, even if it sounds idealistic. Right now the reality of bills and obligations doesn't allow me to teach workshops for free and give away shares of pigs or chickens, but someday I will. To have a goal that makes other people's lives better makes me feel better about people I have hurt, or things that have gone wrong in my past. We all make mistakes, and I don't want to forget mine. I want to balance the cosmic scales of action and consequence however I can.

I couldn't stop thinking about this as Merlin took us along the mountain road home. We passed a woman in her sixties with a shovel trying to reroute some drainage paths. I slowed the horse cart gradually to a stop and asked if she wanted help, either now or later. It would be easy enough to tie Merlin to a tree in their driveway and he would appreciate the break in the shade. The woman thanked us but said she was nearly done and it was easy work. I told her to call or holler anytime, that I live just up the road.

I was feeling a little bit like an ass—sharing stories of altruistic farm ideals, saving elders from their rock piles—and I blushed and felt self-conscious at the polite rebuff. Sara didn't notice; she was just taking in the world from the speed of a trotting horse. As we drove, several other folks stopped by in their truck, letting us go by with a wave or stopping to talk about how nice a day it was. My cart is small, and calling it rickety would be a kindness, but it is safe and pleasant in nice weather. Sara seemed comfortable, even when Merlin stopped to take a dump; his rump was just about two feet from her face. Talk about a chill houseguest. When her husband joined me for a ride later, his first response to moving down the road at a trot was how comfortable it felt. Natural, even. I said horses and humans were partners in transportation for time out of mind before the last hundred years of oil power made automobiles the default. This feels normal because it is. He liked that.

Rides on the horse cart and the Common Sense vegetable harvest were just the first of many activities I was lucky enough to share with them. During the day we also milked the goat and did farm chores. We captured the twin lambs and gave them antitoxin shots and banded their tails. We shot an arrow at a target in the backyard—both of my friends could draw a fifty-pound bow and hit the bull's-eye at ten yards!—and did some good old-fashioned napping and reading under the crimson king maple outside my farm's front door. We ended that long day with our trip to Common Sense. We fell asleep to the sounds of a roaring creek and gentle rain. This all happened on a Tuesday. For that I am grateful enough to fall to my knees and cry.

And there we were at breakfast on a Wednesday. This is a happy place, all this good food around me about to be enjoyed with friends I didn't realize I missed as much as I did. You get so used to living alone, doing the work the farm needs, building your own solitary routine, that you forget the absolute joy of preparing a meal for someone. And when that meal comes from soil and animals in your care or in your community, the joy is amplified in ways so strong and primal we start to understand ourselves as human animals again. A spark ignites, and it lights a fire we didn't realize was aching to burn. This is the work of humankind. It is older and wiser than each of us as an individual. When you give in to it you lose any lingering cynicism and finally can reach out and touch gratitude. When your fingers finally graze it, it is all you can do to not grasp it tight and hold on for dear life.

And when you get it, there's nothing else to do but to dig in.

GREENS

IF FERVENT DESIRE and vitamin D deficiency summoned me into the garden in April, in the heart of June it is all about reaping what I have sown. It's just me out there with a straw hat, a small basket handmade by a friend, and plans for a simple recipe or two. I am neither a skilled gardener nor cook, but I can grow things well enough and prepare them in ways that make me feel satisfied. In a lot of ways, that June garden is a lot like how I play my fiddle. There are six-year-olds with more skill with that instrument, but I was never in it for the stage. I started playing the fiddle to make music that pleases me, calms me, and connects me to places where I once lived and that I miss very much. Playing mountain music from Tennessee or Scottish airs learned in Idaho is enough to make me happy. Tending a patch of kale, potatoes, squash, tomatoes, lettuce, onions, and beans, even if it's weeks before full harvest, is just as rewarding. Sometimes homesteading is all about feeding a pile of friends, potlucks, campfires, and music jams, but sometimes it is just about feeding you. I'm comfortable enough that I can pick something from the earth, cook it up, and sit under the maple tree with my fiddle and play a few songs I know by heart. That beats Julia Child and Juilliard in my book.

I was thinking about the simplest of garden meals: the salad. I began to pick leaves of lettuce, and they soon fill the basket with flecks of dark soil lining their veins. My lettuce patch isn't fancy, more a series of shacks than giant mansions in a row. It never produces big heads but it produces a lot of delicious, ready-to-pick young leaves that make the little raised bed look like a Chia Pet. As my body readjusts to the seasons of all-day outdoor work I find a light lunch better for fueling the day. Here is where those baby greens shine. I grab a fistful or two and run them under the spout of the artesian well on the side of my front yard. I give them a good shake in the air, and sometimes it creates enough of a spray that if the midday sun hits it right, I catch a glimpse of a rainbow's arc. I place the washed greens in a big bowl with some homemade goat cheese, balsamic dressing, and whatever else the garden may have to offer—pea shoots, young herbs. I dig in and eat slowly. I sit on the small deck overlooking the side yard with my feet dangling off the side, and I watch the flock of new layers learn the ropes of chickenhood from the older birds. The chicken coop and barn are just a bow shot away, and from my perch I can watch the birds' dramatic social life as I savor my meal.

Not so long ago I was a much larger woman. I live such an active life that I thought it would be impossible to gain weight, but the truth is, I was constantly consuming food to distract myself from my real problems. See, my first few years of farming were exciting, mind reeling, and exhausting, but they were also extremely lonely, stressful, and depressing. I lacked true self-esteem and let many people take advantage of me. Some took their payment emotionally, others with guilt or judgment. I loved a man whom I let use me as a distraction, too naive and too lonely not to let him. Becoming a farmer and an equestrian made me a stronger woman, and snapped me out of any notion of victimhood or doubt in my abilities. It's hard to let a man walk all over you if you know how to have your way with a thousand pounds of horse.

I have discovered that there is a huge difference between self-confidence and self-esteem. I have ample confidence in my abilities to meet and challenge or conquer any goal, but I think very little of the woman doing those things. I grew up in a house where appearances mattered, and was reminded of it constantly. It was tough being the swarthy, chunky, brunette kid sister to a blonde cheerleader who could date any fellow she wanted. I was a weird kid, an unattractive one, and I knew it. I learned to be funny and found emotional shelter in food. I wanted so much to be thin, to look like the girls on shows of the day such as *Buffy the Vampire Slayer* and *Dawson's Creek,* but knew I wasn't ever going to. This made me dislike myself even more. I saw myself as a failure for not appearing as attractive as the gifts I knew I had inside.

People are often surprised to hear that I lack self-esteem, thinking me a confident person. What they see is a woman who quit her day job and bought her dream farm, all while writing about her life. Accomplishing goals may seem synonymous with satisfaction, but to me the books, job, and farm weren't goals at all. They were things that happened because they had to happen, necessary gains or losses to create space for the life I needed to live. My real goals are so much simpler, and because of that simplicity, it is frustrating as hell when I can't achieve them.

Put simply, I want to accept myself as I am while working toward a healthier body. If there is one place I lack discipline or self-control, it's my diet. I think part of the appeal and initial obsession with homesteading came from a deep-seated desire to escape from the processed foods that have been poisoning me emotionally and physically my entire life. Who wants a donut with chemical frosting when there is a slice of homemade arugula and onion quiche cooling on the counter, made from your own hens' eggs and vegetables? It took a few years, a few pairs of running shoes, and a lot of self-reflection, but as more and more of my meals come from the backyard, less of me wants to pull into drive-thrus for coffee and an éclair. And when the demonic cravings of fat, salt, and sugar overtake me I think about a meal such as a sun-kissed salad, toppling over with goat cheese and dressing, and look down at the slimmer waist I earned from conquering those cravings. . . . Then I stop wanting donuts.

I savor this meal in contemplation and silence—at least I am not speaking, although the homestead is buzzing. I eat outside and enjoy the flavors, oils, and smells of it while the sheep on the hillside contemplate their own simpler life choices. In a few moments the goats will cry to be milked and the chickens will cluck and scatter toward me for feed, but in this stolen moment I let myself eat up a whole chunk of time we call early summer. It is music to my ears.

HARDENING OFF

CHICKENS ARE A LOT LIKE tomato plants. Both need to be started indoors, require care and attention, and taste delicious.

Chickens and tomatoes are also both bad at transitions. You can't just take a hothouse seedling and stick it in your backyard and expect to grow your own Little Italy. Chances are the little buddy will die from the night chill, root damage, or a flat-out refusal to adapt. So how do you ease the young'uns you've started indoors to their own place in a raised bed? You teach them slowly, a technique called hardening off. This is a process in which trays or six-packs of small plants are put out in the sun, wind, and rain during the day and then brought inside at night to the safety of a climate-controlled window side or greenhouse. A few days of this and the plants can stay out past dark and eventually all night without wither, droop, or other cellulose complaints.

The same is true for chickens. You can't take the little chicks from their heat-pumping indoor brooder and stick them in their outdoor pen and expect them to take to it like a full-grown bird. Chicks need their own version of hardening off. I do it with wire cages on sunny days. These cages are the small, movable sort, like a rabbit hutch, with inch-wide mesh that allows access to grass and lets ants and bugs fly through but keeps nosy older chickens, hawks, and neighborhood dogs away. Set out in a sunny spot on even a mild day, these cages give the chicks a chance to feel grass under their feet, scratch in the dirt, peck at bugs, and learn other important chicken life skills slowly. As long as there is a bit of shade, rain, and wind cover (and this can be as simple as a cardboard box turned on its side in the cage) they can be left to explore while I go about weeding and doing chores or other farm gigs. At night when they pile into a fluffy heap to soldier on through to dawn, I scoop them up and bring them back to the warm brooder. As the nights grow warmer and they grow savvier, they gain the ability to handle the outdoors full-time. It's a process, just like bringing up any other living thing, animal or vegetable.

I find that it isn't those first delicate seedlings or the adorable baby animals that draw me into this life as much as the in-between times. Holding a fluffy chick or seeing those green sprouts is encouraging, no doubt, but to me those are just the necessary beginnings. It's the life that is lived among these things, and because of these things, that kindles warm satisfaction. And one of the best examples of this is a little wire cage of fledgling chicks, bopping and chirping in the sunlight as I work around the farm. It doesn't matter if I'm on a business call or knee-deep in goat bedding, cleaning out manure: those sounds ground and center me. More than meditation or a sigh of happy relief those sounds are a reminder of the life I choose to live. I am a woman who, on any given spring day, can turn to the cage of *Gallus gallus domesticus* chicks she is weatherproofing. The cage, as normal a lawn accessory as patio furniture, is now a living part of the scenery, just like the sheep on the hillside and the horses neighing somewhere in the high field. This is my living. To see that cage of birds, or the cold-framed vegetable patch, or even the goat pen out of the corner of my eye is now as normal as seeing a filing cabinet or conference room used to be. I harnessed the mundane and fell ridiculously in love with it.

While chickens may be a more animated subject, the tomatoes are the silent heroes watching all the antics of the feathered set. They climb and twine around their cages and stakes, yellow flowers bursting, pollen storms sweeping through overnight. They are visited by the feral honeybees who used to live in a hive near the garden but have since swarmed off to live in the rickety upper loft of my small barn. (A genius choice on their part, since it isn't safe for people to walk up there, and the pitched roof has become a giant top-bar hive for their personal empire.) It's a dance, this season: the blossoms and the bees and the silly young chickens who think eating a honeybee will make a good snack (a harsh lesson quickly learned). And one day, what had been all yellow flowers are now little bulbs of green, dispersed like grapes every which way. If we are fortunate in rain and sunlight they swell and ripen, and by mid-July I am enjoying those first sun-warmed fruits sliced over basil leaves with a scandalously large slab of soft mozzarella.

It will be a few more weeks, maybe even a month or more, before these small birds are fit enough for the wilds of the barnyard. They will have much to learn from their elders, and the only way for them to get their education is through messy experience. It takes seeing all your peers run like hell under a bush when a hawk soars overheard to take the hint and do the same. They need to follow others to the stream for water, unless a chance whim sends them that way. You can be a rogue chicken if you want, and stay away from the flock. This farm has several beings who prefer the company of the hill sheep or ponies, but the clichés are true: life is easier in a community. And I want not only for these new guys to be used to the layout of the farm but also for all the birds already here to walk up and meet them. It's a lot easier to accept someone who's been watching from the sidelines for a few weeks and has been pecked a bit on the head than to see someone for the first time one fine May day. Chickens are the lot who coined "pecking order," you see. And a blind date with new stock is really just a ringside bell clanging for a battle royal.

WEEDING

THE GARDEN IS A PLACE I GO to when I need to escape, and I mean *really* escape. It may seem more in line with our romantic sensibilities that hopping on the back of a horse and galloping up a mountainside or taking a walk in the woods with the dogs would be better breaks from the stresses of everyday life, but these escapades are nothing like the garden. Because when you garden there is nothing to mind but the static earth and plants, things that do not buck or whinny, chase squirrels, or ask you to throw sticks. In the soil there is just the work of planting, weeding, digging, and hoeing. The type of thinking required for this work is done in the recesses of the brain, and does not involve conscious mental effort. I bring out a radio, stream an audiobook, or blast music; my body is in one place and my mind is in another. I do not think while I pluck stray grass shoots, nettles, and daffodil spikes. I'm lost in a story, singing along with favorite lyrics, or rapt at an episode of *This American Life*. Sometimes I think going into the garden is like stepping into a chamber that transports me to another dimension. It's a place exactly like the one I live in but nothing matters that used to. Arguments with friends, late bills, that overly large mole on my left breast. . . . Things that wake me up at three A.M. most nights are of such little consequence in the garden that a blade of stray grass demands more attention. Mostly because the grass is present at the same moment as my need to remove it, and the remedy is a tangible act I can commit and repeat without any dispute.

I set aside the weeded greens that I can eat or use in salves and tinctures to heal cuts and bruises, or boost my immune system, and place the rest in a bucket for a compost heap deep in the woods. I don't dare just cast them to the side or even let them stir around in the horse manure pile just a few feet away. Weeds are not prisoners of war who can watch their old missions from a cell. They are traitors to the cause and must be taken out back to rot.

With the weeds executed, I return to the garden. Snap peas are winding and blossomed, perfect white flowers that remind me of the dogwoods in Tennessee. Peas are always the first flowers of my gardens. Few summer flavors can compete with the crunch of a fresh pod right off the vine, but it is bittersweet. That reminder of the South is enough to make a woman weep. I was raised and spent all of my adult life through college in Pennsylvania, and that was the emotional and physical landscape I knew. But I moved to Knoxville after graduation, and it was the Volunteer State that hardened me off for adult life. White pea blossoms make me think of spring in the South, and how it felt to be twenty-two years old and starving for a bigger life. Tennessee was my first real love. I'd had boyfriends, sure, but Tennessee taught me to give in to experience and passion with absolute abandon. The state represents exploration, freedom, music, friends, and growing up fast. The romance of leaving the safety of college and family and moving alone to a new region of the country felt like an epic quest out of my fantasy-novel collection. The eighteen-month tornado that followed that move offered the most possibility and personal freedom I had ever had or may have again. I had no family obligations and, besides the two dogs I had adopted as my roommates, no responsibilities. As

I weed my garden bursting with life, I see weeding as a metaphor for the lifelong process of winnowing out that which does not serve me, giving me space for all that feeds me and lets me flourish.

FENCES

Want to know what the boiled-down, reduction-sauce version of absolute frustration is? Get yourself limited pasture, sagging woven wire fences, and smart sheep. I have a small flock of sheep, just six breeding ewes and a few of their lambs and some wool wethers—a dozen critters, counting lambs, and over three acres of fenced paddocks to move them around in. It sounds like a lot of space, and it is, at least for a while. But as the pasture is eaten down and fences start to stretch and sag, what looked in spring like a scene from a coffee table book looks in high summer more like a Depression-era photograph.

All the grass has been eaten down to nubs and moss, so the hillside has lost its ability to retain soil and fine earth washes away in one landslide. What is left is rocky dirt, some dead apple trees, and fences held together with skewed T-posts and baling twine. It's rough. It works part of the time, but it is easy enough for the sheep to escape from and bother the neighbor's lawns. I am constantly torn between ripping all the fencing out and selling the sheep, waiting until I can afford professional fence installation (a ten-thousand-dollar job, at least), and keeping a stiff upper lip and repairing what I can, when I can.

I know farmers and neighbors with livestock who can hand over thousands of dollars for a fence without a second thought. There's nothing wrong with that, and I certainly would do the same if I could, but it's just not a current reality. And when I visit these farms with better fencing I feel that same mix of envy I feel when I hear about people buying a winter's worth of wood or a barn full of hay in one fell swoop. Those are things I struggle to collect all year and still pay the mortgage. If I got rid of every farm animal I owned, it would still be a lot of work paying bills and heating the place, and I can't imagine going through all that without the joy and rewards of my animals. This place has debts and a long road ahead of me to pay them off. I know they will be paid off and then one day I might write a big hay check to fill the barn behind the farmhouse. But for now I am a green-eyed admirer of proper fences and barns full of hay.

I'm not sure urban and suburban folks understand that there is a class system among farmers, and differences as to what people who make a living off their property and livestock use as measures of success. I'm not sure I understand it, either, but around here there are clear differences in what constitutes farmers who have money and farmers who do not. Big, lucrative farms have big, fancy new trucks. They have professional signage and meticulously kept lawns. They have a house with fresh paint, flowers, and clean windows. The kids all have hobbies and sports off the farm, and are driven to riding lessons or football practice. Their fences are straight and good quality. Their barns have solid roofs and clever, efficient feeding and watering systems and planning. A successful farm is a beautiful thing, and when good husbandry with animals and logical planning are part of that beauty, too, there just isn't a more pleasant sight.

Of course, money alone doesn't make the best-tasting egg, prettiest cow, or healthiest rabbit. A place that looks borderline condemned might have the most perfectly planned feeding and watering systems, sleek-coated stock, and healthy human occupants inside. But let's be honest: looks matter. Cold Antler is not a television-ready farm. It's not a country estate turned into a petting zoo. It's a place that I work and depend on for my livelihood, grocery store, entertainment center, gym, and therapist's office. That's a lot of living for one backyard. It shows. And when you make a partial living hosting and teaching workshops and classes there is always the fear that the participants who pay to come here will be disappointed.

My farm is scrappy. The fences are pitiful and the house needs power-washing. The lawn often needs to be mowed and the truck is dented and rusty. I have a college education and make a decent living, but the average drive-by might think otherwise. That's a shame, but what is even more of a shame is that sometimes I let this get to me, too. I see myself through the lens of the disapproving outsider. It's a dangerous place to be.

It's not the physical appearance of those two sheep paddocks that bothers me as much as the effectiveness of the fences. Or the lack of it. It's one thing to have a hidden bad patch, but it's another thing altogether to fail at one of the basic tenets of livestock farming: to keep the livestock *on the farm*. I need better fences, over a larger area, the two pole barns moved to a different spot, and the pasture reseeded before all the topsoil is gone. I am making steps toward doing all this, but the progress is slow. Since the progress is slow, so are the results. It's my own fault and no one else's. A more logical person would keep the sheep flock smaller, add a few chickens or bees, and not even think about things like horses, red tail hawk aviaries, rabbit pens, breeding ducks, trios of geese, woodland pig pens, or tiered hillside gardens. Even if money were no object, that is a lot of management for one person to handle. On bad days I want to rip out my thinning hair and just sell all the animals and let nature take back the fields. At least then it would be green again and neighbors from up the road wouldn't stop by to let me know that one of my lambs is eating their ornamentals.

But here's the thing: This chaos and energy, the food on the hoof—my lambs, my future harvest—that spills out of every loose end, and the people I have met in its creative process are worth every bit of grief. I quit a steady paycheck and health insurance to live here full-time and chase sheep. Maybe many farmers would trade in their good fences for the opportunity to have those very things. In truth, the sheep escapes only happen during a few months of the year, and I am dealing with this problem. And if needing to retrieve a flock of sheep up the road by shaking a bucket of grain in front of me is my biggest livestock complaint, then I know a lot of farmers around here who would slap me for daring to complain about such a silly problem. Heck, part of me loves seeing them run down the paved mountain road to the farmhouse, speckled sunlight on their wooly backs as they baa and trot toward me. It makes me feel like a drover from one of those photos from another time in history, making her way into town with her flock in an obscure Scottish village.

Finding happiness isn't the same as constantly finding things to improve. Fences will be replaced, and when they are the sheep will find another clever way to sneak out. In the meantime, I'll try to be a little less hard on this place's appearance. There's a reason it isn't meticulously maintained: a very full life is buzzing around inside it. Here's to a balance between home repairs and home-cooked meals.

WINDFALL

THE APPLES WERE GREEN AND DENSE, hanging off the trees so thick that the branches sank from the weight. It was a happy scene for this home brewer, especially after the prior year's late-spring frost that froze the blossoms right off the trees. That was a sad season, for none of the local apples came into fruition. But this year the orchard was making up for it. Those trees were groaning, and in a few months their apples would be ready to be shaken down with my hooked shepherds' crook. You spread a drop cloth beneath the tree, then slip the crook over a branch and give it some rough shakes. The apples all slam down onto the drop cloth, which you bundle up like an old-timey robber from a silent film.

The apple trees are in the middle of a sheep paddock, and the sheep cannot tolerate this theft. I keep them penned away from the tree I am harvesting and they bawl and cry out, incredulous and pissed that I dare take their apples away from them. As far as they are concerned, everything in the pasture is theirs, and every time a strong wind blows, they all run to the base of the trees to snatch up any green-apple drops. It's a comic endeavor, watching them slurp up the apples. Sheep do not have upper teeth, just a hard plate their bottom teeth grind against to masticate grass. So when something as big and juicy as an apple needs to be dealt with, their eyes grow big, their mouths open as wide as they can stretch them, and they squish them with one hard crush against their upper mouth. It is the exact sound that an apple makes when we cider makers throw one into the grinder above the pressing bucket.

This competition for fruit starts out friendly as the summer goes from the solstice into the dog days. A strong wind blows and I look out the kitchen window and smile as a pile of woolies attack the green drops. When Sal, my biggest sheep and favorite wether, eats these he closes his eyes and lifts up his head so none of the juice drips out onto the ground. It may be the most contented look any living creature on earth has ever shown me. But as the days creep toward fall, the game is *on*. No longer are those apples for the indulgent ovine hoards. I have been known, on occasion, to race out to the paddock and beat the sheep to the windfall, or to pen them up first with a bribe of sweet feed when I know a stormy day is coming close to cidering day. Every apple is a part of our harvest, and when you finally uncork those bottles of dry, potent scrumpy, you are validated in your sheep-thwarting efforts with every sip.

BLESSED HUMIDITY

IN JULY MY WORLD TURNS LIQUID. When I step out my front door with Gibson to do the morning chores I enter a brand-new world from the one I had left the night before. It had been so dry, for so long, and now the blessed humidity is back and I am bubbling with energy. My body instantly bursts into a light sheen of sweat. I take in a deep breath of the wet air and let it fill my lungs, smiling. In half an hour of chores my body will be dripping, and then I change into running gear and really learn what humidity is. Water and life are everywhere. I am so in love with humidity.

People complain about this weather because they feel it is uncomfortable. Why is being uncomfortable bad if you are a healthy, young, human animal? Why do we think constant comfort is normal or good? To me, humidity is the reason I am comfortable. Sweat and heat, those are not things to avoid. Those are the things that make my body hum. One July night I sat outside in a hammock in the dark, watching thousands of fireflies light up my farm while thunder rumbled in the distance and lightning danced. The storm was far off. If I was in danger, I didn't care and wasn't moving from under the giant broad-leafed crimson king maple. I was barefoot, and below my swinging feet was a lawn of soft grass and clover. My arms were feeling long and strong. They were deep brown in their summer tan, and wet with sweat (a default condition). Bits of hay and chaff covered them like sparkles, sticking to my skin like so many set jewels. I felt lush and heaving and alive out there swaying above the world. I felt young and full of possibilities and I could not stop smiling.

All around me the fireflies waltzed, the thunder purred, and the sky was shot with light. I was so tired from a day of shooting, riding, writing, and running, and this resting pose felt decadent. Even though the world was hot, I knew what it felt like to be dripping sweat and gasping for breath as I ran up my mountain to drop pounds and remove toxins from my body. In comparison this humid night was a cool breeze blowing through a sauna. It was heavenly.

Humidity provides. Humidity means I live in a place where water is so abundant that it thrives in the very air, in the mud under my feet, in the flashing sky. My friend Othniel says we live in a deciduous rain forest, and he is right. The Northeast is so lush, so alive, that you can put a rock out on your front lawn in the shade and it will grow moss from nutrients summoned like a spell from air, minerals, and time. Life is always finding a way here.

This thunderstorm was about to hit, and I waited. What used to be a reason to run and hide was pure anticipation now. I am not foolish enough to tempt lightning, but I wanted to sate myself in wind and light rain before heading for cover. Just as the seasons of the year have a perfect cycle stretched out over a year of holidays, traditions, birth, life, and harvest, the water cycle runs its course in a matter of hours. Humidity hits its glass ceiling and the atmosphere explodes into a rainstorm. Rain from the storm soaks into the thirsty earth, creating the vegetables, the grasses, the thick green leaves on the bowing trees that whip and dance in the fury of the high winds. What doesn't get taken by plants and animals sinks down into the water table, slides into streams, ponds, and reservoirs. Here it continues to make things thrive. Like the Wheel of the Year, a thunderstorm is a perfect circle. It's a holy moment in time, as precious as the life of a kale leaf or newborn lamb. Water is life on this little planet, and I no longer feel the desire to avoid it in any way. I feel better when immersed in it, dripping sweat and ruining my hair. A humid morning means this part of the world stays healthy, hearts keep pumping, and the long ride into August becomes all the more a reason to get wet.

The Battenkill is a trout river so clean and cold that you can see the bottom six feet below you. At any moment you may share the river with otters, beavers, ducks, and cedar waxwings. I know friends who, while duck hunting in the dead of winter, have watched otters swirl past them; or they have rounded a corner in a quiet canoe to find a pair of bobcats sunning on a rock. This is a wild river, and a boat-launch dock is just four miles or so from my front door. It takes just moments to hop into the truck and dive in for a swim. There's an old Scottish saying that the cure is more of the same. I find this is true about humidity. If you are miserable in the muggy, soppy air, then you might as well just give in and immerse yourself in liquid. When I do this the volume of everything is turned down. What was unbearable heat is now perfect.

A COMPASSIONATE HARVEST

THERE ARE FEW THINGS more satisfying on a small farm than seeing the strut of a fat, full-grown American Bresse cockerel on a summer afternoon. All through the spring and summer these birds gain weight and size, and now, in mid-July, they are at harvest weight. The breed is sturdy and athletic (as far as chickens go), standing tall on their signature large blue legs. Their bodies are packed with a large breast, strong thighs, and petite wings. If you placed a small watermelon on a pair of sturdy stilts and gave it the personality to walk down a runway, you'll have an idea what kind of chickens I am raising. Large red combs flop over the sides of their faces, reminding me of bangs on a male model. Pair that cocky strut with a waving hairdo and perfect feathers and you have an animal any small holder would be proud to say she raised.

The birds are completely oblivious of my smiling approval, though. All day they worship the sun, putting up with it as long as they can and then forming their little party of explorers to raid the streamsides for grubs or to check out their girls by the feeder. Now at full maturity they seem more interested in the hens that were raised alongside them, who are just as lovely as the gents but lack their confident stride, deep chests, taller stance, and fancy combs. I love watching all of these birds together, gathering at the feeder or moving as one unit of bright feathers across the yard to the well to drink their fill. These are happy birds, that's for certain. Their lives are full of food and sex, open ground and sunlight. They have never been restricted to a coop, and it shows—in their savvy caution regarding high-soaring hawks and neighbors' dogs trotting along the road. They think, and they eat their fill, and they flee danger and crave one another's companionship. It's quite the little society. Sometimes I go outside with a glass of ice water or a beer at the end of the day and sit down under the king maple just to watch them.

I've raised other breeds of meat birds before, but nothing like these fellows. The Cornish Cross and Colored Ranger birds I have brought up here had the same rate of growth and the same access to the farm, but they just didn't have the Bresse's style. The American Bresse is a fancy breed, a variation on the French Bresse line famous for its flavor. The bird is so sought after that people have been known to pay eighty euros for one without a second thought, and swear that they are worth every penny. Mine came from a barter deal with a professional breeder. Just watching them parade around the farm, I am pretty sure they know their reputation. They aren't like other chickens around here. These birds shine.

Their feathers are white, and when the sun hits them I swear they glow. When I see these handsome creatures out on the green patches of grass, basking in the sun with outstretched wings, I think of them loading up on sunlight like stick-on glow-in-the-dark stars. At night, when I do the last round of animal checks before bedtime, I look in on them in their coop, snugly situated on their roost, and I am always disappointed that they aren't luminescent.

I used to do all the work of harvesting the birds here on the farm, but now I have the help of a farmer a few towns over. Ben Shaw and his family run their own pastured poultry enterprise and offer all the small farms around here their services in slaughtering the animal humanely, eviscerating it, and packaging the meat. If I only have a trio of birds to prepare, I still do the work in the backyard, but usually I harvest between ten and fifteen birds at a time. If I have anything over half a dozen I bring them in my pickup truck to Ben's place, a twenty-minute drive from Cold Antler. For $3.75 a bird he will accept a crate of my live chickens and with his staff's help and professional equipment have the meat cleaned and bagged and ready for me to pick up in my large cooler in under an hour.

To be perfectly honest, I am grateful for someone else to perform the task. Out of all the animals raised for food, I find chickens to be the most time-consuming to prepare to eat. Using calming inversion cones, supersharp knives for quick, humane deaths (the birds' throats are slit), temperature-controlled hot-water baths to loosen feathers, and special rubber-finger plucking devices, Ben and his crew can have all my birds ready for me. I just have to write a check and drive them home to their new coop in the freezer.

To some this may sound harsh, or even heartless. At one time it did to me as well, but after a few years of raising chickens for food, the work of preparing them to be cooked and eaten has lost all its unsettling aspects. I think it is the loss of a life, not given up freely, that upsets some people. After all, the Bresses did not ask to be killed, nor did they have any idea while they were dancing under the light of the noonday sun that their days were numbered. But we need to understand that livestock brought into this world by human beings, such as meat chickens, are not victims of violence. They are nourishment, and killed not wantonly but out of a deep gratitude. There isn't a human being alive who can live without taking another life. Sometimes that life is a plant's, and sometimes it's an animal's, but both are living things that will no longer grow, reproduce, or thrive among the windrows and under full moons once we consume them. My birds live a great life, have a swift humane death, and are prepared with all the care and joy I can muster. It's an honor having them be a part of my life, and my diet, and to share them with friends and family who feel the same way.

At age twenty I became a vegetarian for animal rights reasons, and spent ten years as one. It was a long road to again think of other sentient beings as food, with a lot of internal struggle. But as a lover of animals I realized the world was not black and white, not as simple as I wanted it to be. I wanted to believe that not eating animals meant less animal suffering, but that wasn't the case. This position may suit some people's diet or spirituality, but it felt like pacifism to me. As an activist, I petitioned, shopped, and ate in a way that I felt was helping animals. But my choice to avoid leather and hotdogs was not helping animals waiting in slaughter pens, it was just taking me out of the meat-purchasing market entirely. My vote as a consumer to impact how these poor animals lived was removed by that choice to opt out of eating meat. There isn't a single person in an industrial meat company boardroom that cares about the sales of veggie burgers. But those same people quiver when they see the scale shift from factory-farmed meat to grass-fed or free-range. The meat-eating public, the quality of meat consumed, and the way it is consumed are changing how the public feels about their product. Just as Upton Sinclair's *The Jungle* rocked the meatpacking industry back in the early twentieth century, the farmers who raise meat humanely, CSA members, and other patrons are a voice for compassion and kindness to the living things we take as food.

That doesn't mean that choosing to avoid animal flesh isn't still a compassionate choice. It does take you out of the argument and reduce the number of human beings eating the ridiculous amount of meat per head that we consume. I am not invalidating that choice, or any choice, to eat consciously. Personally I chose to be a part of the small group of people taking animals to grass, letting them live full lives, and giving people an option that doesn't come packed in Styrofoam and displayed under fluorescent lights, along with a chance to connect with their food animals in a meaningful and respectful way. I depend on my pigs, sheep, goats, chickens, ducks, rabbits, and turkeys to feed me and my community. I depend on my goats for milk and my chickens for eggs, too. It feels correct to my animal self, my soul, and the traditions I come from. I eat animals because I care too much about them not to.

So I raise these beautiful animals who are born to die. I watch these chickens and feel nothing but safety, gratitude, and a deep-kindred appreciation for all that they are. I give them a place to grow and glow and then ask one last favor, their lives in return. Some of the hens will stay to create a breeding program to keep the Bresses going strong, but most will provide meals and stories. They will be remembered and appreciated with every bite. And every bird I drive across town lines to Ben Shaw's farm keeps one more factory farm bird off a consumer's table. It supports a small business, an alternative to mass production, and I can enjoy that famous flavor with a clean conscience.

The Perfect Bird

THE BEST ROAST CHICKEN YOU'LL EVER EAT!

If there's one kitchen trick everyone should learn, it's how to roast a chicken. Roasting a chicken is such a satisfying, savory, home-warming skill, and a roasted three- to five-pound bird can furnish three to six meals for an outlay of just a couple of dollars in flesh, potatoes, and carrots. The way I do it is easy and inexpensive, and it always turns out wonderful. Prep time is just minutes, and you only need one pan, a bowl, and a knife to cut veggies with. Follow these instructions and I promise you, you'll want to roast a bird every chance you get. If you bought the chicken from the store, or if it was recently frozen, brining is the way to ensure that your roasted chicken is moist and savory, instead of stringy and dry. I use an adaptation from Hugh Fearnley-Whittingstall's *River Cottage Meat Book,* with brining options I learned from the Cooks Illustrated website.

You'll need, to roast:

ONE SMALL ROASTING CHICKEN (3–5 POUNDS)

OLIVE OIL (OR WARMED BUTTER)

CRUSHED HERBS (FINELY CHOPPED GARLIC, SAGE, COARSE SALT, AND ROSEMARY) OR A COMMERCIAL CHICKEN MEAT RUB

PIECE OF ALUMINUM FOIL

ROASTING PAN

MEAT THERMOMETER

3–4 MEDIUM POTATOES

4–6 CARROTS

You'll need, for brining (optional):

LARGE BOWL, SAUCEPAN, OR ONE-GALLON ZIPLOC-TYPE FREEZER BAG

⅔ CUP SALT

¾ CUP SUGAR

SPRIG OF ROSEMARY

A FEW BAY LEAVES

To brine, take your whole bird and place it in a large bowl, saucepan, or Ziploc bag with enough water to cover the bird. Add the salt, sugar, rosemary, and bay leaves. Let this set in the fridge for two to four hours, turning the bird every hour or so. When you're ready to cook it, it'll be primed.

Preheat the oven to 420 degrees.

Take your fresh, defrosted-in-the-fridge, or brined chicken and rinse it in cold water. I rinse out the cavity, under the wings, everything, to reduce saltiness, but it is optional. Give it a few good shakes in the sink. Put it into a large bowl and set it aside. Take out your roasting pan—I use a glass Pyrex pan. Cut up carrots and potatoes into chunks no larger than your thumb and cover the bottom of the roasting pan with them. Besides cooking in the bird's juices and fats, they act as a roasting rack, letting air under your bird and helping it cook thoroughly. I always brush a light coating of olive oil and chicken rub spices over my veggies as well, but you don't have to. Set this aside and go back to your bird-in-bowl.

Take soft salted butter or olive oil and rub the entire bird over with it. With the bird chest-up, take a knife and slice the bird's breast skin. Try to get your fingers right under the breast skin and slide butter or oil into it, right over the breast meat. If the idea of an inner-skin massage makes you want to gag, use a knife to make some cuts and slide some fat under the breast skin to allow air and steam to get between the skin and the muscles. This method keeps the skin browned and the breast moist. Trust me, this step is worth the trouble.

Last, take the crushed herbs or commercial meat rub and coat your bird entirely with this wonderful mix. If you want, tie the drumsticks together with some butcher's string (at your kitchen store). Place the bird on top of your cut veggies. Stick your meat thermometer into the thickest part of the breast.

Now, open that oven door, baby. Slide your herb-rubbed chicken into the hot oven. The cooking method is a flash of heat followed by a slower roast. Let the bird crackle and pop in there for twenty to thirty minutes. Then lower the heat to 350 degrees and cover the bird with a shield of aluminum foil lightly placed over it. This stops the skin from scorching but allows it to get a little crispy. I roast the bird for at least an hour at the lower heat, taking it out when the bird is a nice brown color to check the temperature. If your meat thermometer reads 170 degrees in the thickest part of the breast, the bird should be cooked through. Stab the bird's skin to check that the juices run clear, not milky or red. If you wiggle the legs they should be almost ready to come right off in your hand. If your bird seems to have a lower temperature, just pop it back in for twenty minutes and check again later.

If all the signs are good, let the bird sit for twenty minutes— the meat will keep cooking as the bird cools on your stove top— then serve it with a side of savory carrots and taters! Enjoy!

HOMESTEADER'S VACATION

BY MID-JULY THIS FARM IS SINGING. The chaos of spring is long past. The lambs, the kids, the young pigs, and chickens have all outgrown that time of dire need and attention and the days feel long as weeks. Every farm has a lull period, and it differs for each farmer, depending on what she produces and the schedule she keeps. Here at Cold Antler I am lulled by the weeks between the summer solstice and the first of August, around six weeks of blessed routine and nothing else. I wake up and do all the normal chores, and by midmorning the milking things are washed, the garden is watered and weeded, and I am a free woman until evening chores. If I wanted to I could drive my truck down to Albany, hop on a train, and be in the middle of New York City by one P.M., walk around Central Park, or check out the newest exhibit at the MoMA, and be home in time for evening milking, with hours until dark. Up until just a few years ago, that would have sounded like heaven. To wake up on your own farm and have a full summer day free to travel and explore a wild city, and then come home to the comfort of your own croft. Amazing, right? But the woman I have become could not imagine such a trip, and not because it doesn't sound enjoyable or realistic, but because the concept of distance has changed for me. An hour's drive on a highway followed by two hours in a fast train may just be a quarter turn of the clock face, but there is a decadence to it now that unsettles me. If I want some time in a city I drive the lazy forty-five minutes to downtown Saratoga, or perhaps Glens Falls. These are far smaller towns, but when you live on a mountain with sheep, any place that offers Korean barbecue seems exotic.

I don't dislike the city. Quite the opposite. I miss it. Part of me feels a dull ache for it. But New York City seems much farther away than three hours. The paradigm change is powerful. I chose this life of chicks and milking goats and leaving behind any semblance of my old life, and part of that concerns the practice of travel. I don't go anywhere. It's impossible with the amount of livestock and responsibilities on the farm, but above that I don't want to leave. It's not agoraphobia, it's just contentment. There's so much going on here, and even more happening on the friends' farms around me. Within ten miles of my place are activities, work, friends, food, and entertainment beyond measure. Three miles from my front door is the town of Cambridge, where I have seen plays, ballets, poetry slams, and world-renowned musicians traveling to our opera house. A few doors down from Hubbard Hall is Battenkill Books, which hosts everything from book clubs to local history lectures to knitting circles to book signings with acclaimed authors. There are cafés and little diners with local and hearty fare. The old train depot whence freight used to be shipped down the Hudson River to the city is now a community arts center hosting everything from bluegrass jams to Irish step-dancing classes. Cambridge has a farmers' market, several bars, more churches, a saddle club, and I'm sure many events I haven't even been told about. It keeps me busy and enlightened to more culture than a trio of turkeys in the tall grass at home can offer.

Plus, I know the people, and they know me. I can walk into

the hardware store and pick up sheep feed and everyone knows my name and says hello to Gibson, who comes wherever I go. Small-town establishments aren't worried about sheepdogs biting customers and being sued. It's a casual, artistic, and hardworking town with a tray of dog biscuits at the bank teller's window.

On a summer day it is rare for me even to make it as far as Cambridge because that means getting into the truck and leaving the farm. There is so much to keep me busy just outside the front door. There is always more weeding to do, and plugging a pair of speakers into the iPod lets me get lost in an audiobook with my bare feet on the dark earth, just barely missing stepping on the fat toads that hop about the lush grass. Next to the gardens are rounds of ash and locust, a gift from a friend who works on a road crew who cuts down trees on public roadsides. The trees' fate is to lie by the side of the road and rot. To someone who heats entirely with firewood this is road kill of the best sort and should not be wasted. My friend hauled it here in his little Chevy pickup and together we unloaded it. I split the ash—which splits so fine you feel twice as strong.

Beside the fire pile is the pony cart my friends and I just repainted and fixed up. We kicked out the old rotten wooden seats and back and added new pine boards, stained and everything. I can hook up little Jasper to it and get in a training session, trotting down the winding road in a horse cart and getting lost in the poetry of it. Just beyond the horse cart is my backyard archery range. Every day I practice with my fifty-pound draw recurve. I bought it last summer from Joseph the Bold, a man of local legend in traditional archery circles. It is lined with leather and sinew, and has a beautiful leather grip and a soft tanned skin arrow rest. But what I will choose over any trip to town, bow shot, weed patch, or ax is my saddle and Merlin. It only takes me about half an hour to gather, groom, and tack up the gentle beast, and then I am trotting across the street to my neighbor's trails. There we can take full advantage of a snowmobiler and ATV lover's handiwork and explore forest, field, and mountain stream whenever we want. And I do. And when I come home and curry Merlin's wet coat, I kiss that black horse on the forehead and walk him by the lead rope to his back pasture to play with Jasper in the fresh new grass. With the horses content there is a hammock under the maple tree in front of the house, just above the burbling lullaby of the artesian well, and there I will grab a book and a glass of water or wine, and read.

I can vacation right outside my front door.

I have friends who work very lucrative jobs at impressive companies, or used to have them, at least. The social entropy that happens when you choose a totally different lifestyle than your peers (a few states away, to boot) makes for some pretty wide chasms. I don't think any of my old friends harbor any ill will toward me or mine, but it's hard for them to have a normal conversation when all a person has to talk about is the last week of haying. I gained a farm, but I lost some friends.

I find that I don't like that about myself, this single-mindedness I get over whatever agricultural fanfare is going on. It's like inviting someone to a dinner party and the guest only talks about one thing all night. That's fine if it's a dinner party at a Star Trek convention, but my old college friends can only listen to so much talk about breeding rabbits and sheep sheds. And my single-mindedness in working toward my goal made me too blinded by excitement to realize what was happening. I suppose that happens to a lot of inexperienced girls in love. You can't see far beyond the chemistry, and growing my own food and the homestead around it was the thing that made me fall more intensely in love with life than anything before.

So although there are museums, concerts, and vacations out there that I would love to experience with old friends, the reality is that none of that is a part of my life anymore, not really. A day trip to a museum means at least five to ten hours away from my home. That means leaving dogs indoors without bathroom breaks for five to twelve hours. That means calling a friend to come into your home to walk said dogs. It means a goat with a full udder of milk is praying you don't get stuck in traffic or miss the bus. It means leaving detailed instructions with neighbors on what to do if the sheep escape and how not to get electrocuted by the fence. It means no one refills empty water tanks, no one is there for animal emergencies and lambing snafus. Or, it means having a roommate or spouse who can do all these things for you. I'm single and raise livestock of several varieties near a wood-heated home. I'm not going anywhere. If I needed to travel I'd grow just vegetables with irrigation lines and good fences. But carrots don't take you at a canter up a mountainside at sunset. Those carrots are what I feed my farm animals for snacks.

So how does one avoid becoming a hermit? Simple. Friends and family come see me. I host potlucks, game nights, workshops, and friends from out of town. We entertain ourselves right here in the country, on the farm or close to it. The more adventurous can hike the mountain, swim the river, fish the streams, or join me in a cart ride with a pony to visit friends. We head into town for lunch at a café, buy a book, shop the antique stores, or drive through this beautiful county. We cook good food at each other's farms that we enjoy with cold, tall glasses of adult beverages. We do all the things I used to watch people in those farm magazines doing, but we're not posing at all. We are communing.

I also make time to visit friends on their farms, see what everyone else is working on between whatever it is they are working on.

One year I made a trip on one of the hottest days of the summer to the Daughton family's Firecracker Farm. It was a partial visit, partial favor. My truck's "check-engine" light went on, and knowing as much about the workings of motor vehicles as I do about thoracic surgery, I asked if Tim Daughton could take a look at it. He obliged but said to come for dinner as well if I was hungry; his wife had a grill feast ready to go. So I drove down with Gibson and I caught up with his wife, Cathy, and their sons while Tim hooked up my truck's computer to a code reader. It turned out I had a loose gas cap, just a twenty-two-dollar repair. I was so happy I hugged him. Engine trouble is no blessing on a shoestring budget. Feeling like I had dodged the worst, I was elated and asked to look around their farm. So much had happened there since I last visited in late winter. Seeing it would be a treat.

Cathy and Tim walked me around their land and barn. Their gardens were thriving, their poultry quirky and beautiful as ever. They now had two goats in their barn that came from my farm. I had traded my smaller milking doe, Francis, and a kid of that spring's trio of goatlings for help with some construction work later in the summer. Francis and her son, who was named Nacho, were looking sleek-coated and happy. We scratched ears and talked about converting their goat pens back into pig pens and then headed down to the field where they had raised meat birds the previous summer. My eyes grew three sizes too large as I stared, mouth open and smile wide, at the field. It was full of wheat! The Daughtons laughed and said it was an accident. Turned out the movable chicken tractors they dragged around the field to take meat chickens to new pasture were more than just a single operation. Everywhere a tractor had been set down now had stalks of wheat growing tall with big fat seed heads blowing gently in the evening breeze.

Tim laughed and explained that he guessed the chickens didn't like the wheat in their scratch grains! I laughed too because it was a perfect example of sustainable, smart farming, and it had happened by accident. They didn't realize the scratching and pooping birds in those little pens were cultivating and fertilizing what would become next year's flour. Amazed I walked up to the long-stemmed wheat and looked closer at a seed head. I quickly realized I knew nothing about wheat at all. I had some general knowledge that the grains needed to be gathered, dried, winnowed (separating chaff from the grains), and then milled into flour. But that was pretty basic. How did you know when the grain was ready to harvest? How long does it dry? What kind of storage problems can you have? What do you grind it with? I knew chickens inside and out and could make you a serious tomato sauce or apple pie from my own backyard, but when it comes to something like harvesting grains I am as ignorant as a bag of horseshoes.

Cathy explained Wheat 101 to me. She grabbed a seed head and pulled it apart to reveal the tiny, pealike grains. She took her fingernail and dented it. "When the grain is soft enough to dent but too hard to pop, it's called dough stage. Soon it will be hard enough to barely scratch and that is when you cut it from the field and hang it to dry." I took one of the little white pearls in her hand and prodded it with a fingernail. It dented slightly but it wasn't a balloon. She said this was the later dough stage. I tried to puzzle out in my head whether she meant bread dough or the female deer. I didn't say anything, feeling foolish around this woman who had learned enough about the grain to know when to hang it and dry it. She went on to tell me that before this dough stage was a stage called the milk stage—when the grains are sweet as candy and bursting with a white liquid. The wheat goes from milk to dough to hardening off to what is considered ripe. When it's ripe it loses any green tint and turns a homely brown. I was entranced by all this, looking at the seeds in my rough palm and trying to picture their life story from chicken coop poo pile to a loaf of artisan bread.

I adore bread, all bread, and have never raised my own grains to bake an extremely local source of it. Cathy and Tim were planning on grinding a little of the seed, but using most of it as seed for the following spring. I drove home that night with visions of brick backyard bread ovens and pizza parties by a campfire. I already grew my own tomato sauce and made my own cheese. Could growing grain be so hard? I mean, the Daughtons had done it by accident and it worked out. Would they rent me their chickens?

Bread is such a wholesome, beautiful, blessed food. It's been unfairly demonized in our current society, and I find it disturbing that entire diets are based on avoiding all carbohydrates. To label any homegrown food bad because it causes us to gain weight is truly a first-world problem. There is so much food around right now that we can pick on who is better than who, create fad diets, and replace ingredients to accommodate whatever product craze crops up. Eating too much of anything is bad, and eating enough pasta in one sitting to power medieval field workers isn't a smart choice when all you are doing is sitting at a computer. But that doesn't make wheat bad, it just means our use of it is uninformed at times. And we still have an instinctive desire to pack away energy, even when we know the commute home is an air-conditioned sedan ride and not a fourteen-mile hike.

We can all find a balance with bread in a way that makes sense to us. Learning the stages of wheat seed-head growth by name felt like unlocking a secret code, lost lore. It was the key to making the best bread in the world. In my downtime between seasons, that quick visit to a surprise wheat field made me want to set aside the novels and streamside hooch of my homesteader's vacation and start digging up clay for a beehive brick oven. If wheat could be grown in a raised bed, sign me up. I have nothing to lose, and I may even score some dough.

Ridiculously Easy Crusty Bread

This is the easiest bread recipe I can offer you—even easier than the highly popular no-knead Dutch oven recipes. Anyone can do it, even if you have never, ever baked bread from scratch before. You don't need anything but a mixing bowl, flour, water, salt, active dry yeast, and some sort of round bakeware to let it rise and bake in. It is an overnight, no-knead rise, so it's not "insta-bread," but for about five minutes of effort before work in the morning you can have amazingly fresh, crusty bread every night for dinner. This is rustic bread, almost peasant food by today's standards. But I can think of nothing cozier or more comforting after a chilly day of farm chores than warm, homemade bread—unfussy and practical.

This is an adaption from the no-knead recipe featured in the *New York Times* awhile back.

1½ CUPS HOT (NOT BOILING) WATER

1 TEASPOON ACTIVE DRY YEAST

3 CUPS FLOUR, HALF WHOLE-WHEAT AND HALF WHITE

1 TEASPOON SALT

CAST-IRON SKILLET, DUTCH OVEN, OR PYREX CAKE PAN

1. Pour the hot water into a mixing bowl and add the active dry yeast. Let the mixture set for five minutes. When the water is

cloudy and an active foam is bubbling on the top, your yeast is activated and you are ready for step 2.

2. Mix the flour one cup at a time into the yeasty warm water. Add the salt while you turn the water into a sticky, even paste free of lumps.

3. Cover it with a damp cloth and leave it to sit at room temperature for 8 to 12 hours.

4. When you return after your day at work, or after a night of sleep, check the bread dough. It should be bubbly and doubled in size. It is ready!

5. Sprinkle flour on your table, flour your hands well, and take out the whole doughy mass. Fold it over itself a few times and make it into a ball. You'll need flour on your hands to stop it from sticking. Now gently place the ball in your bowl again and let it rise for another hour or two.

6. Twenty minutes before you intend to start baking the bread, start preheating the oven to 450 degrees. Put the skillet or baking pan into the oven when you start the preheating. It has to heat up with the oven.

7. When the temperature reaches 450, take the baking pan out and place your ball of dough in it. It doesn't matter if it loses its shape—it'll bake evenly.

8. Bake at 450 until the bread is browned, 30 to 45 minutes. Keep an eye on it.

9. When it is done, take it out and let it cool for at least half an hour before slicing. Enjoy!

AUTUMN

HARVEST

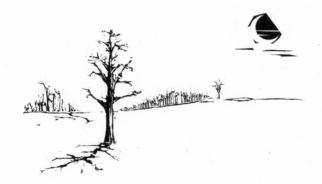

LUGHNASA

THE LIGHT CHANGES FIRST. Upon stepping outside the farmhouse in early August, you can't help but notice this. The harshness and power that made morning sun in June and July seem foreboding has faded. There isn't the mugginess of humidity or that hum of electricity that swirls around early July mornings and requires you never to leave home without a change of clothes, a gallon of water, a towel for the river, and enough sunscreen to suffocate a whippet. The day may still grow hot in August, but just as the light has matured, so has the heat. There isn't the same ferociousness to it. Or perhaps, from working outside all summer, I've grown accustomed to it. In August I could work in long sleeves and long pants all day if I preferred. In July that kind of wardrobe only gets donned for haying; anyone else that covered up would feel they were voluntarily smothering themselves.

The urgency has faded from the light and I can tell that it is tired. At midday the light feels as if it is near dusk; it's stretched thin. The first fallen leaves cast shadows across the lawn at sunrise. And those leaves on the grass aren't the storm orphans you may see after high winds pass through; they have begun to change color and fall into senescence. I never noticed how early this change happened before I started homesteading. In all my childhood memories, one day the leaves changed colors, and never in those long late-August days. I didn't think they were allowed to change before school started and I was toting a respectable supply of notebooks and a new backpack. Now, in August when I look outside I notice that the staghorn sumac has entirely red branches; it's a tree on fire compared to the locusts behind it. Up along the forest's edge I can see other trees start to drop leaves as well. The old apple trees, some maples that get the most sun. Our modern culture says this time of year is all about beach balls and vacation days, but on a small farm this far north I see through the merchandising of summer.

The light's loss of urgency signals a new urgency on my part. I must think of setting by supplies to provide for food, hay, and heat for the coming winter. Long before I showed up in this world, people all over Europe (my ancestors—perhaps yours, too) knew about this change in the light. And it caused a lot more concern than it does for us modern rural residents. Like me they were farmers, albeit at a totally different level of urgency and self-sufficiency. If my canned goods run out or my freezer shorts out I may be out some money and inconvenienced, but the grocery store is waiting for me to swipe my plastic card and restock. But it wasn't that long ago that the food you started collecting in August and kept collecting through the rest of the harvest season had to keep you alive until spring. August was a time to work like crazy, harvest, and prepare for winter. It wasn't all drudgery, either.

The Wheel of the Year that I celebrate has three harvest festivals. The first one starts around August 2. People in the past knew the severe consequences of not harvesting and storing enough food for the winter, but they also knew when to have a good time. The three months of harvest—August, September, and October—are marked by special holidays dedicated to food and life. It all starts in August, with Lughnasa.

Lughnasa has been celebrated for a long time. A Celtic holiday, it was observed widely around the British Isles up until a couple of centuries ago. Its pre-Christian origins are clear in its name, which comes from the god of light, Lugh in ancient Celtic. Today Lughnasa is also the modern Gaelic word for the month of August. So this is a word and holiday that has been around a long time, but started to fade into obscurity as industry ramped up. Lughnasa was a festival of gratitude for the first harvests, usually grains, berries, and early tree fruits, among other local foragings. Being the first of three holy harvest days, it was usually celebrated with festivals, feasts, and games. Some celebrants made pilgrimages to holy sites or wells, asking for blessings and luck for the dark months ahead. The closest modern, urban-dweller equivalent might be packing a new hoodie you ordered for a trail hike in case it got a little nippy at night. But anyone going for a brisk walk and prepared for a change in the weather is unknowingly celebrating Lughnasa, to my way of thinking. It's a day dedicated to physical effort, good food, and preparation for and acceptance of change.

My Lughnasa celebrations rarely are observances in any traditional or ritual sense. There are no Beltane bonfires to leap over, potlucks planned, or caber-tossing happening at Cold Antler. The closest thing I may be able to make time for is canning with some friends in my kitchen. August here is just as busy as April, but instead of preparing for the months of births, planting, and growth, I need to start collecting all those stores and getting them just as ready for winter as I am getting myself. There is jam to make, sauce to can, peppers and onions to dice and freeze, and potatoes to pull and stick in a cool place. I don't grow enough to feed myself from my garden all winter (yet!), so I also get great deals on farmers' market seconds. Farmers still need to get rid of the bruised and ugly fruits and veggies that don't have the looks to sell on market day. Otherwise they'll just turn into compost. I can get a case of bruised tomatoes at a deep discount at the farm stand down the road in August. That makes a lot of pint jars of sauce. I may spend a whole day with a marinara pot simmering on the stove, followed by canning in the water bath. It really does eat up a whole afternoon and evening. But to see those jars on the pantry shelf and taste that same August-kitchen smell when I pop open the jar and pour some sauce over homemade egg noodles on a snowy day have me praising Lugh in every sense. God of light, indeed. He manages to bring back August when the woodstove is fighting a minus-five-degrees wind-chill factor. Bless him.

Now, I can't think about August without thinking about autumn. My farm's world is changing, and colder weather will come faster than I realize. By September I am waking up to mornings when I can see my breath, and by October we get snowfalls. I have a lot to prepare between now and October if I want to get through the winter comfortably. There is still firewood to split and stack in the weatherproof shed near my kitchen. If a friend or two show up with chain saws and axes we can produce a cord a day, turning big tree sections into smaller rounds, and splitting those with heavy mauls to be stored for my two stoves. I still have an oil furnace, but only because the hot water system is attached to it while I save for a solar version. I used to buy heating oil by the hundred-gallon load. I now depend mostly on my stoves. For the summer a five-gallon container keeps me in hot water for two weeks, but by winter I'll need at least fifty gallons in store. The temperature in early August is likely to be in the eighties, yet all I can think about is firewood and hot water on a future February morning. Well, that and hay.

Hay is a constant need on this farm. All around my mountain

are fields of grass that could be turned into fine hay, but those neighbors don't have animals, so they don't harvest their hay, and I don't have the equipment to harvest it myself. I need to gather and store enough pasture for a long, upstate winter. Hay hoarding begins in early August. I put up at least twenty bales a week in the barn, in addition to my regular supply, which is stacked by the side of the house and kept dry under a tarp. I feed out three to four bales a day from the tarp hay to the sheep, goats, and horses, but I do not touch the bales in the barn. They do not exist in my mind as an available resource until the first snowfall comes down heavy. If I run out of my regular supply of hay, I can resort to the barn hay. I'm lucky that within ten miles of my place I have two hay sources, and within twenty miles, countless. By late October I want to have at least two hundred bales laying in wait for when they are needed. That should last until Christmas. I restock hay as it is used, replacing twenty or thirty bales at a time. This is sustainable for my animals and my cash flow.

I also try to put up as much of my own food as possible. At any given moment Cold Antler Farm has between two and six months' worth of food on hand. Some of it is stored in cans, jars, and cupboards, but most of it is out in the backyard. It's in my chickens, goats, and pigs in the form of eggs, milk, and meat. It's in the vegetable garden soil, too. Between stored food, living and working homestead systems, and emergency rations I am as well stocked for my own winter meals as the sheep and horses are with their hay. That may sound like the talk of a backwoods survivalist—and in a sense, it is. I am a single woman who lives alone on a farm and has an unpredictable income in an already shaky economy. I'm not preparing for the end of the world, but I am preparing for times when things get uncomfortable. I have learned the hard way what it is like on those long stretches between book deals when the income I make from workshops and speaking events needs to cover essentials like the mortgage and car payment. Grocery shopping isn't in that thin budget, much less eating out, and to be able to go months without spending a penny on food is a huge safety net and a weight lifted from my shoulders. As someone who has nearly had her sole vehicle repossessed and been threatened by banks with foreclosure, I've been scared straight into making sure that the animals and I have basic needs covered when extra cash is thin.

So by late summer all of this squirrel-like gathering starts to stir in me. I have slowly learned this mind-set via trial and error, some long power outages, and the reality of my personal economy. If I wanted to live in relative peace and comfort as a farmer on a mountain I needed to have things like dependable heat and food for myself and all I care for. My first winter in the house I dealt with some blackouts from winter storms, and since my heating system was entirely based on electricity I spent a few very, very uncomfortable days in the dark. I also had months when I had all the electricity I could afford to use but the price of heating oil was so high that one month with the thermostat set in the low fifties cost over $400. The following summer I installed a woodstove for heating and cooking in the center of my living room. That and a second, larger woodstove in my mudroom keeps the house toasty and off of the grid. The best part? The fuel grows all over this mountain and right on my own property.

If it weren't for this farm I'm not sure I would ever have cared about heating, hay, and food storage. My whole life it was just buttons and dials on appliances, grocery stores, and paychecks. I felt that as long as I had a job from which I could live paycheck to paycheck, I was as secure as anyone can be in modern America.

But spending one winter alone on a mountainside with livestock changed that view. I used to think an electronically deposited check and a 401(k) meant security. The first thing I did when I cashed out my little 401(k) was buy that woodstove and have it professionally installed. It's the best four grand I ever spent! Security isn't a paycheck. It's knowing that when your world shuts off on you in the middle of a blizzard, you and yours will not be cold, hungry, scared, or uncomfortable. Because I have this level of self-reliance I feel incredibly safe, and that matters to me. Lugh's blessing, and my own luck, carry me through.

DINNER DATE

I HAVE A RULE that I never write about my personal life, certainly not my dates, for the public. I respect both my partner's and my own right to privacy. So that part of me isn't on display or up for discussion. But some dinner dates are so great, *so breathtaking,* that I feel obligated, as a writer and artist, to share them with the world.

I was taken out to dinner recently by a guy I've been with on and off for a while. He's smart and tall, and has hair that you lose yourself in if you let yourself touch it. Dark brown eyes and a knowing grin. A smart-ass through and through, just the way I like them. He's from the UK, and while he's not a big talker, he has a way about him that intrigues the hell out of me. I fell for him hard, at first sight, and since he walked into my life it hasn't been the same. I wish the same for every woman, to know this kind of satisfaction, companionship, and use of her thigh muscles. . . .

I had a dinner date with my Fell pony. Merlin took me out to dinner. It was that golden afternoon light, right before sunset, and Merlin was tied up to the hitching post outside my little farmhouse. He carried saddlebags laden with a cooler of drinks and food, utensils, napkins, and an icepack as well as the gear always on hand during a trail ride: a spare halter, lead rope, hoof pick, bug spray, and such. Tied to the top of the saddlebags, behind the saddle, was a wool blanket my mother had given me. I used baling twine to lash it down. This was my kind of night out. I couldn't hold back my grin as the golden rays split through the trees. I hopped onto Merlin's back and we reined out onto the road at a trot.

We rode to the dirt path that leads to our trailhead. Once there, we walked, trotted, and cantered across field, forest, and stream. Merlin stopped to drink and splash and I stretched, feeling the heat of the welcome sun on my face, knowing the season in a way few lucky people do. We walked along a running stream, up a steep little cliffside as we headed toward an open hayfield. Once there I hopped off, tied Merlin by a lead top and halter to a small tree, and unloaded the blanket, dinner, and e-reader for my respite. This would be a night to remember.

The meal wasn't fancy, but it was delicious. I ate pasta I had packed and I sipped cider as I sat on the blanket. I read a chapter from S. M. Stirling's *The Protector's War* as I munched. The weather was a perfect seventy degrees. The kind of seventy degrees only people who spent the winter feeding a woodstove on minus-thirteen-degree nights can appreciate. I savored the food. I sipped the cider like it was nectar from heaven. I read and laughed quietly to myself, listening to the tail swishes and occasional sighs from Merlin.

To some people eating out means going to a restaurant. To me "out" means being outside and so wrapped up in the moment that I lose track of time. I spent an hour out there in the grass, reading and occasionally talking to a horse. He was good company and seemed to appreciate the peppermint treats I had in my pocket and the cold stream water he got to slurp and splash his heavy feet in.

We stayed outside until the sun had left my eastern side of the mountain and a light chill blew across the field. I had a light sweater in the saddlebags and slid it over my head. I thought about how in a few weeks I'd be a shell of my former self, exhausted from an afternoon of helping neighbors with haying and stacking bales. As I secured the gear and talked softly to Merlin, I wondered to myself what a moonlit walk home on those trails among the creeks and fireflies would be like. How it would soak into my skin and become a memory I told people about for the rest of my life. When I grow nostalgic at something that hasn't even happened yet, I know it is time to go home. I packed up the rest of the picnic into Merlin's saddlebags and tied off the blanket roll and we headed back down the mountain to Cold Antler. My stomach was full, my horse content, and the road short, with chores like milking and chick feeding ahead. Work never stops here, but little vacations like this balance out the anxieties.

SEARCHING FOR SHIPS

THE MIND SEES WHAT IT WANTS TO SEE, what it is trained to see, and what it expects to see.

I have always been fascinated by a tale about Columbus's ships: Natives watching the sea from the shoreline couldn't see the billowing sails and wooden vessels until they were a couple hundred yards from shore because the human mind can't see what its brain has no prior knowledge or conception of. The brain has to process the information and get it to the eyes. The more unexpected the concept is, the longer this takes. So the ships were cognitively invisible, inconceivable to those without the slightest notion of what European merchant vessels under sail looked like. Similarly, Mesoamericans, at first sight of a horse and rider, thought it was one animal instead of two, like a centaur.

One August night, I was enjoying a cool end to a very hot day. The afternoon had been a scorcher, exactly the kind I prefer a summer day to be. It was in the nineties and humid. I do not air-condition my house. I do not try to avoid heat or discomfort. I embrace it. I love the way it forces me to sweat and move like an animal instead of some doughy frump in a morgue. I like the way I can feel beads of sweat leave my brow and coat my back as a result of my doing normal chores. I can feel the water weight and toxins leaving my body. I swill water and spend as much time in the sun as

I can, loving the soupy air, the dark greenness of it all. Right now the mountains are alive. Every rock is growing moss, every plant is dripping dew, and every young animal born in April has learned to lope across forest and hillside alike. It is something to behold, this wet and happy summer.

I had spent the day farming and gardening. I milked the goat and shot my daily quota of arrows. When I got to the point of feeling dizzy I rested in the shade of the maple with a glass of cool water and my unshod feet dangling into the little pool by the well. By sunset the lack of sun and a light wind had me in a sweatshirt, even though it was still eighty degrees. My body was so adapted to and accepting of the discomfort of heat that the lack of it gave me a chill. So in a sweater and kilt I swayed in my hammock, not thirty feet from the bubbling creek that runs down my mountain road and through my farm. I was reading *The Fellowship of the Ring,* and stayed out well into dark. A good book can entrance you like that, making hours swirl around you until they are gone.

I spend most of my time alone. I have friends and neighbors whom I visit, or they come here. I go to classes and local events. Anyone who would call me a hermit would be sorely mistaken. But compared to how most people live, rural or urban, I spend a lot of time alone. I've been single since freshman year of college, well over a decade ago. That's been my choice as much as not, since I have no interest in dating for sport or company. I would only want to be involved with someone if I fell in love, which I have done a couple of times, but the guys I was so fond of were not interested. I would rather be single than involved with someone I wasn't crazy about, mostly because you have to be crazy to put up with another animal as erratic and frustrating as a human being. Horses, sheep, dogs, and goats—despite all their quirks—are predictable and dependable. But humans are fickle and contrary, confusing and hurtful, and I am certain I am just as bad as anyone else. So until love, the real thing, comes along, I am happy to be authentically alone rather than in a contrived relationship.

I know a few couples who are genuinely happy, but most people I meet are still together out of habit, insecurity, reputation, or the obligation to prior investments. I'm not sure I believe in ideas like marriage as a pathway to happiness. For every happy married couple there seems to be another fifty that see it as a trap they can't escape. I am terrified of getting myself in that trap, or putting someone else in it. People always tell me how dangerous it is to ride Merlin without a helmet, or to live without health insurance, but what seems really dangerous is the slow grind of hopelessness from being stuck in a sad relationship. So I am only in this game for keeps, and only with someone I could not imagine living without.

Living alone can be lonely, so I sleep in a double bed with a border collie. I would like to think that that will change, but if it doesn't that is okay too. The farm, the humid days, the nights like this where absolute contentedness comes only from the physical work and the escape of a good book—I have fallen in love with my own life. It's enough for this tired woman. But if the right man came along I'd buy a dog bed in a second.

While reading in the dark, by the glow of the e-reader's backlit screen, I thought I caught a flash in the corner of my eye. I looked and saw nothing, and disregarded it as glare from my glasses. Then another flash caught my peripheral vision, and then another. I turned to look at the darkest, wettest, green-black swirl of forest above the stream. I felt like the native looking for the outline of a ship, trying to remember the pace and flash of the holy glow in the distance. My mind strained to piece it together, to decipher the timing. There was nothing but the rush of my heartbeat and flailing memory. Suddenly I thought of catching fireflies in glass pickle jars with holes punched into their metal lids. I thought of the drive-in movies, and how you knew that when the flashes arrived, the picture was about to begin. I thought of staring at these fireflies from a hammock in Vermont, and from a hidden riverside cabin in Tennessee. I thought of how I missed them so much when I lived in the Northwest; their absence over those dry summers felt incorrect, naturally wrong. Fireflies are grounding agents in my life, which seems peaceful and bucolic to some, because I live on a farm. My life is fraught with just as much confusion and anxiety as any modern person's. Fireflies remind me of beautiful times and places; I rely on them to transport me. I thought of all of this, and tried to see the horse and rider through the darkness as two animals. Thunder rumbled in the distance, a common and gentle sound around these parts and my heart ripped right open. The lights were not mysteries at all.

And then again I saw a flash.

And then I saw five more.

The farm was alive with hundreds of fireflies.

There are places you can go where you can escape discomfort. There are places you can plug in machines to pretend weather doesn't exist. There are people who will tell you that humidity is a horrible thing and should be avoided at all costs. Do not go to those places, avoid plugs when you can, and never believe a liar because there is nothing more beautiful in this angry, scary world than a hundred fireflies in the dark of a lightning-kissed sky. Nothing.

MABON

WHEN I WAS SEVENTEEN YEARS OLD I asked my boyfriend to join me for a very special occasion: the autumnal equinox, a holiday for his earthy girlfriend. Although he didn't share my beliefs in the Wheel of the Year—my leanings have been rooted for years—he was game for a picnic. I wanted to take him to a spot I often hiked to on the Appalachian Trail, an overlook of the green valley where we lived that was also a popular camping spot for thru hikers. We'd also see the burst of color from the fall foliage. It wasn't the peak of color in the mid-Atlantic forests, but it was close. I knew this trail well, and we hiked in the late afternoon, setting ourselves up to have our picnic at dusk. Once there I made a fire in the charcoal pit surrounded by heavy stones and started cooking our dinner. I had packed everything I needed for a quick meal, and after we ate we sipped some wine pilfered from my parents and snuggled on the mountainside, overlooking the beauty all around us. "What do you call this day, again?" he asked me, hugging me from behind. "Mabon" was my happy reply. I remember taking in long, deep breaths of forest, woodsmoke, and the gentle decay of early fall. I was in my favorite clothing in the world at the time, jeans and a hooded sweatshirt, and felt part of something deeply traditional and magical. "Mabon. Best day of the year."

Mabon is the second of the three harvest festivals celebrated on the Wheel of the Year. It's the heart of the harvest. Lughnasa is the official opening of ceremonies, and Mabon is the bright and breathing spectacle of gratitude. It is shameless in its bounty, the sun is still shining, and food is rolling in. Vegetables, meat, eggs, milk, grains, hay for the animals, and more are coming off our fields and pastures at an alarming rate. Even the backyard garden is starting to let some of the Sun Gold tomatoes on the vines rot before they can be canned up or added to the pasta sauce pot. Mabon has been known as Harvest Home; as wine harvest—in some areas of Europe they are bringing in the grapes as we are bringing in the corn; and by an older Gaelic name, Meán Fómhair. It was a time of work as much as play, a celebration and necessary break from farm and field to sit back with a glass of cider or wine and take in the wealth, the genuine wealth, all around you.

Some people cannot get enough of the pomp and celebration between Thanksgiving and Christmas. It fills them up with vibrating anticipation for the big day. They have a list of traditions or rituals they can depend on, things they know will be shared with loved ones. They get the Christmas tree from the same tree farm on the second Sunday of December with Uncle Mike. The town caroling starts at Mrs. Gilmore's house at six P.M. on the winter solstice. Yuletide is a big deal, and I love it, too, but for me it doesn't have the verve of fall. My window of anticipation is that exciting time between the autumnal equinox and Halloween, what I call Mabon and Samhain.

Mabon is my true Thanksgiving, the peak of the harvest around Cold Antler. This is the time when I am fervently preserving the garden's bounty and am filling the freezers with chickens, rabbits, and turkeys—and hopefully, someday, some venison, too. I am buying fifty-pound sacks of potatoes from the local farm stands for less than twenty dollars apiece, and storing them with my own potatoes for winter meals. The staggering abundance means that if you know how to sweet-talk a little you can leave your local farmers' market with a criminal amount of tomatoes for canning, dehydrating, or freezing. The official holiday for food-related gratitude, a Thursday in late November, is not a date set by farmers.

Between the richness of food everywhere I look and the glory of a firecracker fall on the horizon I am beside myself at Mabon. The weather is perfect: still warm enough to work up a sweat on a trail ride at noon but campfire-cold at night. The sky seems bluer, the green things darker, and like any end-of-year party Mabon has the energy of last-chance celebration. I'm a hunter, and the deer are twitterpated as all get out, happy to rut and mate and act with total abandon throughout my county's fields and forests. Turkeys are strutting their stuff, too. Smaller game like pheasants, rabbits, and squirrels chatter and seem to be everywhere you look. Few people nowadays can experience the wild abundance of nature paired with the domesticity of farms. City folk and those who live in more arid regions or in built-up suburbia can't eat their fill of locally grown tomatoes and potatoes and locally shot venison steaks whenever they want. But here in Washington County it is our normal. I feel lucky and blessed to have landed here. If the computer-run world crashed to a halt, here in the upper Hudson Valley—at least those of us paying attention to horny deer and deals on sacks of spuds— we would be worried about how to store all the food, not how we were going to eat it.

In my own farm traditions, Mabon is synonymous with hard cider. Around this time every fall my friends and I gather to collect apples, press them, and start brewing our hooch. What started with a small gathering a few years ago is now an all-out potluck celebration where everyone brings his or her favorite comfort foods. Friends and fellow homesteaders Chrissy and Tyler bring her mac and cheese, I bring pulled pork (soaked in apple cider), and James always brings some beer. It's a work party, so we socialize as we powerwash fruit, grind apples, and take turns pressing the juice. But by the end of the day we have full stomachs, a light buzz, and enough gallons of fresh cider to drown in. The usual take is twenty to thirty gallons, stored in five-gallon fermenting buckets or carboys. It's the last celebration of the apples I have been watching since the sheep were sheared back in the spring and the apple blossoms were sending me into a hopeful reverie. And by Mabon we are all wearing wool sweaters and beanies as we take turns making juice out of what was once just a promise from a blossom. You may not believe in religion, or the power of prayer, or anything else that science hasn't given much clout to. But anyone who has gotten drunk on New Year's Eve from the juice of the fruit of a spring flower knows that magic happens. You just need to work for it a little.

How to Make a Small Batch of Hard Cider

Making a small batch of hard cider is a great no-fear way to get into home brewing, and it's a great way to support your local orchards, too. For a small batch you need very few supplies, and you will get nearly two gallons of the good ol' scrumpy that will be ready to bottle for the winter holidays! Not a bad way to show up at a Christmas party.

Because this cider is made with honey, it is called *ciser,* the traditional term for honey-based cider.

To make hard cider you need freshly pressed apple cider, yeast, and honey. You want the kind of cider that has no additives in it and no "nutrition facts" on the label. The best place to get it is from a local orchard where they press their own apples and sell the juice from their farm. Around here, it is everywhere. But even if you live in an urban area I'm sure the orchards ship it to local co-ops, natural food stores, and farmers' markets. Just make sure that what you are buying is plain apple cider, nothing fancy.

To turn that cider into alcohol you just need a few tools that any online brew shop, such as Northern Brewer, can ship out to you, or there may be a brew shop near you.

First, you need a small two-gallon fermentation bucket with a lid that has a grommeted airlock hole in it. I suggest buying one from the pros, as they aren't expensive and you are certain to get a fresh and airtight seal. Northern Brewer sells these for a few bucks. You also need an airlock. This is a small, usually plastic device that is fitted into a plastic lid with a rubber grommet or into a carboy bung that acts as a one-way valve during fermentation. Airlocks allow the pressure built up within your fermenter to escape without allowing outside air to get back in. The third "tool" you need is a sanitizer to sanitize your equipment and eliminate all unwanted yeasts and so forth. I use Star San Sanitizer. You also need a special yeast, called champagne yeast. You can get all of these tools and supplies online or at your local brew shop.

Now, let's make cider! You need:

2-GALLON FERMENTATION BUCKET

AIRLOCK

STAR SAN SANITIZER

½ PACKET OF CHAMPAGNE YEAST (SUCH AS RED STAR
  PASTEUR CHAMPAGNE)

1½ POUNDS OF HONEY

1. Set out your cider on the counter to let it come up to room temperature before brewing. The process is faster and the yeast is more active if the cider isn't cold.

2. Sanitize you bucket and airlock (just throw it in the bucket). Fill the bucket nearly full with clean tap water and add a little more than ¼ oz. of Star San (following manufacturer's instructions). Cover the bucket and seal the lid tight. Cover the grommeted hole with your finger and shake the bucket a little to make sure that the sanitizer makes contact with all parts of the inside lid and the sides of the bucket. When done, pour out the liquid (it will be foamy, but foam is okay) and set aside. Do *not* rinse with more tap water. Remove the airlock and set it aside on a clean plate.

3. Pour in the fresh cider. Add the honey. No need to stir.

4. Pour in half a package of champagne dry yeast. No need to stir that either.

5. Place lid on tight. Check all around to make sure that the seal is good.

6. Insert airlock in lid. Make sure that this seal is also good.

7. Set in a dark, quiet place to ferment and bubble.

It really is that simple. You can make it more complicated if you like and heat up the cider first and stir in warm honey and so on. Mixing ingredients will make it ferment faster, but I am all about spending as little time as possible brewing and more time farming. In about a day or two you will see bubbling coming out of your airlock. That means it is working! Right there in your own home or cabinet you are creating an alcoholic drink, and not just any alcohol but really, really good apple cider. When bubbling stops (two weeks to a month later) let the cider remain in the same place at least another week. The yeast will settle, and then you can siphon the cider into sanitized glass bottles with lids. At this point it is ready to drink, but I like to let it season a bit longer. It sits in dark green or brown beer bottles or wine bottles in a cabinet until I am ready to pour it out and enjoy it. But be mindful and responsible. Homebrew cider is usually around 12 to 15 percent alcohol.

SWEATER SONG

I raise sheep for meat, but so far, in my own kitchen, that intention has only been theoretical. I have bartered lambs for things the farm needed such as firewood, lumber, and help with building projects, and I have eaten meat from lambs that I raised but that other people have slaughtered and prepared for the table. But I have never been present at the death of one of my sheep, nor have I ever had their meat stored in my freezer. I don't have any qualms with the idea of doing so—it just hasn't happened yet. Lambs are a currency of high value around here, and folks with a firewood surplus and lamb deficit have made me some handsome trades. It was better for the entire farm to wait for my own lamb masala and crowns of ribs in favor of getting the barn fixed and having fires to heat the place through winter.

Instead of the meat, I am focused on my sheep's wool. It is a good economical choice, too. Sure, a sheep can produce up to eighty pounds of mutton, but mutton is a one-shot deal. Wool is a renewable resource. And I do know my way around a fleece. Give me a five-pound armload of hay-and-sheep-poop-covered wool and I know how to wash it, dry it, card it, spin it, and knit it into a jaunty hat.

Sheep's wool is a grass-fed renewable resource. It is harvested once a year, close to lambing season in the spring, and packed into large, brown wardrobe boxes. It's shipped to a small wool mill in

Connecticut that turns it into yarn, felt, and wool rugs. I have one of the two-by-three-foot rugs in my bathroom, and I mailed my sister the second—the only other one in existence—as a housewarming present. It may have been just another rug to her, but that bath mat really is the story of an entire year on a mountainside farm. That rough wool from a flock of rare Scottish and English sheep was tended to every day by her littler sister, the long locks cut and turned into a home good few people can find constructed locally.

My flock doesn't just keep my feet warm in the loo. The wool from the two types of sheep have different purposes. The meat sheep with their long waterproof coats provide beautiful rugs and layers of soft felt, but the Longwools have decadent curly locks. My Cotswolds, Border Leicesters, and Romneys have soft fleeces that anyone would like to feel against their skin. There is nothing itchy about them. When their coats are shorn and the fleeces picked clean and washed, they have an almost pearl-like glow to them, a shine with a slightly golden halo that never appears dirty. When it is professionally cleaned, carded, and spun into skeins of wool ready for knitting needles, it is precious beyond understanding. And when yarn that luminescent is turned into a hat or scarf, it brings more than satisfaction. It brings a high, and then you are hooked.

Not only do the sheep offer me this wool, but they eat, sleep, and play outside my farmhouse's door along with me all year long. It is a partnership. On any cold fall day you slide a handknit wool cap down over your ears and lyrics from an entire year of sheep song are singing in your brain. And suddenly that little homespun hat has become a talisman. It draws good feelings about yourself into you, showing you exactly what you are capable of.

When I step outside for late September chores with a homemade hat on my head, hauling hay or pouring chicken feed into a big top-loading feeder aren't even chores anymore. *What? You need breakfast? Don't worry, I have this covered, for all of you little chicks. Of course I can offer you a second helping. Can't you see I created warmth out of thin air?*

Making a thing is special. Making a thing that covers one of your basic needs for survival is magic.

Making Yarn from Raw Wool

Turning wool into yarn is not a complicated process. It's a lot easier than turning yarn into socks or sweaters. Knitting remains a popular craft, but few knitters can look at a sheep and go through the steps to turn out yarn that is needle-worthy. When I teach wool workshops here at the farm I always start in the field, with a pair of shears. Showing people the rawness at the start makes the final product seem like something from Rumpelstiltskin's spinning wheel.

SKIRTING AND WASHING

Skirting is the first step in cleaning raw wool. It is done to remove the biggest chunks of debris from the wool. Take a small amount of raw wool—what would comfortably fit squeezed inside a quart mason jar—and set it down on a flat outdoor surface such as a picnic table. Pull out any pieces of hay or grass, little clumps of sheep dung, or clumps of dirt that you see. What you pull off is almost always organic material, so it can usually go right into your compost pile.

Once skirted set the wool aside and get a small basin of water and some dish detergent. The basin can be galvanized steel in the form of a livestock trough from the feed store or a rubber storage container. All it needs to do is hold water and be light enough for you to pick it up and dump. Fill this vessel halfway full of cold water and add about a tablespoon of detergent. Let it disperse naturally or sink to the bottom. Do not swish it around to make suds. Now, add your clump of skirted wool and gently press it down until it is submerged. Do not agitate it, or wash it like you would a pair of underwear or socks. Any sort of fussing will turn the wool into a ball of felt. Just let it soak. I do this either outdoors or in my shower stall, a place where I can easily let it soak without bother and dump the water out easily.

Let it soak as long as you like. Some folks say it has to be soaked overnight and some give it five minutes. I usually let that first soak last an hour and then gently remove the entire clump of wet wool and set it on a clean grill rack. These are inexpensive and found at most hardware stories, but old screens for windows work just as well. The water will be dirty as all get out, usually a dull and almost opaque gray. Dump this water, rinse out your basin, and repeat the process for another soak. Do this as many times as it takes for the water to pour out clear. When that happens your wool is as clean as it is going to be and is ready to be dried in the sun. Set it back on your grill or screen and put it in the full sun, in a spot where it won't be bothered by chickens, dogs, or toddlers. Elevate it so that air can circulate all around it. In a few hours it will be beautiful, fluffy, clean wool. Congratulations, you just processed wool off a sheep. Feels great, doesn't it?

CARDING

Carding means taking the clean wool and forcing the fibers to stretch and unkink from their swirls and curls. Carding brushes are the most common tools used to do this by hand. They look like large, wooden dog slicker brushes with row after row of metal teeth. A more efficient carding tool is a drum carder, a round teeth-lined barrel that constantly pulls at the wool as you crank the handle. Both are great tools for home processing, but the hand carders are far more affordable. Drum carders can cost hundreds of dollars. A pair of slicker brushes might be waiting for you at the dollar store.

Take your hand cards (or big slicker brushes) and drag them together so their teeth pull against each other. I always pull my cards so the handles face away from each other on opposite sides, creating maximum Velcro-like pull. This is the carding way! Place a small amount of clean wool on the surface of one carder and brush the carders back and forth against each other, pulling the fibers straight. This starts to create long, bulky skeins of wool called roving, in which the fibers are pretty well aligned. Roving is then spun into yarn. In yarn shops and at wool festivals roving is most usually seen in the form of big, thick, wooly rolls of fiber, about an inch in diameter, rolled up into what looks like a yarn ball for a tiger-size cat to play with. But home carding creates rolags, straight little candy-bar sized piles of wool ready to be spun by more primitive means. Once you have a few rolags you are ready to spin!

SPINNING WITH A DROP SPINDLE

The easiest and cheapest way to spin is to use a drop spindle, the original spinning device that predates wheels and machines. It's a small tool, usually under a foot long, and consists of a round wooden wheel and a dowel. The wheel is called the whorl and the dowel is called the shaft. The round whorl creates the weight and something to twirl that will twist your yarn onto the shaft. It is so easy—it is just spinning a toy that twists your pieces of rolag into yarn. You can buy beautiful handcrafted and mahogany-stained drop spindles or make them out of an old CD and a dowel from the craft store. My first drop spindle was the CD type, given to me at a heritage days festival in my hometown. I learned on it, and now, fifteen years later, I live with sheep. So take that as fair warning.

To begin you need a leader around eighteen inches long. This is any piece of yarn you have lying around the house. Tie it to the bottom of your drop spindle, under the whorl on the small piece of the shaft. Bring it up and over the whorl on the opposite side with the longer shaft that ends with a small metal hook. Twist the leader around the shaft at the whorl's base a few times and then snag it to the metal hook. You are ready to spin!

Fray the end of your leader yarn a bit. Take a section of your rolag and attach some loose fibers of wool to the frayed end of your leader yarn. With your left hand pinch these together and hold them connected as you suspend the drop spindle in the air below it. Give it a spin with your right hand in a clockwise motion and watch as the loose strands of the rolag wool start to twist onto the leader. You just connected your yarn to the leader and are officially spinning! Let a little more wool out of your left hand and see it gather in the twist you are making with the constant swirl of the spindle. As you do this your drop spindle sinks more and more to the floor as your yarn strand grows longer. When it "drops" and hits the floor, stop and wind your new yarn around the whorl. Leave a little fluff at the end of your new hand-spun leader end and you are ready to do it again. Congratulations! You just created a single-ply yarn. You did it with some dirty wool, tap water, soap, dog brushes, and a CD on a stick. Who knew?!

HUNTING

I LOVE ANIMALS and I love the hunt—every aspect of it. I love getting up early in the morning and packing my shotgun and digging out my father's orange jacket. I love the hugs from friends who arrive with white clouds of breath escaping with their words. I love the excited whimpers of the dogs and the way a mug of coffee feels in my hands on the truck ride to the state lands. I love the anticipation—by far my favorite drug. And most of all I love the brisk pace you keep up behind a dog with a nose in the thick brush, and the burst of energy as a cock bird is flushed up into the sky. There are a few seconds of communication: Who's taking the shot? Where did the bird drop? But mostly the hunt comprises moments of pure adrenaline between intense and tempered strides across the landscape.

I was with Holden Daughton, Patty Wesner, and Patty's dog Harley. Harley's a Large Münsterländer and a hell of a tracker. With that dog we had the secret weapon against the hiding birds. He would sniff them out and scare them up into the air where we could take safe shots with our guns. I had my twelve gauge with upland and small-game shots. Patty and Holden both had the lighter twenty gauge shotguns better suited to bird hunting. But my trusty pump Mossberg is my all-around gun. I use it for turkeys, pheasants, and varmints, too.

We walked across the fields and wetlands for hours, in and out of rain showers, watching the dog and smiling wide as we got a chance to take home a pheasant each for our farms. Patty got her bird first, a fat hen. I got to shoot at my bird second, a nice flush and straight line of flight only ten yards away from me. I managed to just hit him in the bum, but he went down, and Harley helped us find him when he did. After those two back-to-back successes it took a long time to find Holden's bird. We had just about given up and were practically back to the truck when the biggest cock bird we had seen all day shot up into the air, and Holden smote it down. Harley retrieved it from the tree line and we three happy hunters went back the truck with grins across our tired faces. We had walked for hours, stood in the chill rain, and had maintained the level of constant alertness that had made just a few hours in the woods feel like a marathon.

To some my dual love for animals and hunting sounds like a cringing contradiction, and I understand that. My place in the scheme of things is as a pack animal that hunts by daylight, where my bliss writhes and turns up to the sun.

For all the birds, so far my hunting trips have not produced a single ounce of venison for the table. As a new hunter I have been given the gift of several chances to take some beautiful deer, but my inexperience, hesitation, and general clumsiness have had a way of trumping any opportunities granted. It seems like the deer have caught on to our intentions, and spots once crawling with cervine activity are now barren as the harvested cornfields around Washington County. Well, barren of corn. Seems like every harvested cornfield in the county is crawling with deer I can't legally shoot. Deer are like men. You see them everywhere, but that doesn't mean you can have them.

But during hunting season I am trying, boy am I trying. I bow hunt with my fifty-pound recurve, a traditional style of bow without a scope, bells, wheels, or whistles. When bow season ends I use the hunting rifle my father gave to me. I'm the middle kid and he has a son, but neither of my siblings hunts, or wants to. So the beautiful heirloom was handed to me. In the deep fall, if I am not taking care of the farm or running errands, I am in the forest or on a tree stand. The hunt has taken on a mythical veil, a wild and sacred thing. It is more than just aiming a gun at a buck. It is hours and hours of silent meditation, but meditation on the edge. Like sitting in the lotus position on the edge of your roof. Probably nothing will happen, but if it does you'd better be ready, safe, fast, and wolf-quick in your decisions. It's exhausting and frustrating and exhilarating at the same time.

Most of hunting means just sitting, just waiting. Friends and fellow hunters have all sorts of words of advice on which animals to take and which to avoid. I listen but think to myself that it won't matter to me if it's a ten-point buck or a graying doe on the lam from another hunter. I will take the animal that chance, luck, and a good quick death offer (if I am lucky enough to have such a chance). If I do manage to shoot, kill, and gut a deer it will be genuinely thanked and honored for offering its life. It will be professionally butchered. It will feed myself and friends for months over storied meals of how the hunt went down.

This outlook—to approach the hunt with respect, patience, and wonder—is what makes me a hunter, not the actual taking of a life. To me a hunter is someone who takes life for the table, not the wall mount. She takes it with humility and the understanding that we, too, will die someday.

I remember finding these words of Kahil Gibran in Barbara

Kingsolver's homesteading memoir, *Animal, Vegetable, Miracle: A Year of Food Life,* and they have stayed with me: "When you kill a beast say to him in your heart, 'By the same power that slays you, I too am slain; and I too shall be consumed. For the law that delivered you into my hand shall deliver me into a mightier hand. Your blood and my blood is naught but the sap that feeds the tree of heaven.'"

So although I have not managed to take a deer into the mightier hand, here is what I have tucked away in a game bag close to my heart:

I have sat for hours in the forest and remembered again what a joy it is to simply sit still. I have snuck up on a great blue heron, and looked at its offended eyes before it flew away with wildly loud flaps of its wings. I have shared a tree stand with a chickadee, singing inches from my ears. I know what the sound of a flock of geese sounds like overhead when it's not honking. Their wing strokes are sirens at sunset. I have seen bucks trot, antlers raised to attention, and does coyly avoid my virginal hands as they take aim. I've sat through snowflakes, and sunrises, and watched a baby fawn cry for its mother as it ran across a field in the cold blue dawn. I laughed with crows. I studied owl songs. I stared at tracks and heard stories of a dozen hunters and their hunts. I have done all this, and hope to until I grow too feeble to pull a bowstring.

I am joining a sisterhood and brotherhood of people who have reconnected with a primal urge. Not to kill wildlife but to provide for their loved ones and family in the most basic way possible: with deeply nutritious food. It is not a sport of death, not really. It's a sport for survivors: the brave, the patient, the storytellers.

HALLOWMAS

Hallowmas has been the anchor of agrarians for thousands of years. It is the oldest continuously celebrated religious holiday we have. Its history is either lost, flawed, candy-coated (literally), or despised by those who see it as demonic and a sacrilege. In truth it is none of these things. It's not a time for costume parties or a ticket to eternal damnation. It's a reminder that you are still among the living, and that makes you one of the lucky ones. It reminds us to be joyous and grateful while remembering the people we have lost since the previous October. I'm honored to be able to introduce you to my Holiest Time of the Year. But please let me cover a bit of history so we're all on the same page.

The name Halloween, as most people know, is verbal shorthand for All Hallow's Eve. The word "eve" is a shortening of "evening." We're used to hearing the more popular Christmas Eve, as a term for the day before a big holiday. All Hallows' Eve is the night before the Catholic celebration of All Saints' Day on November 1 (those who wear halos are "hallowed"). All Saints' Day was always supposed to be the big show, but we can all agree that that has changed. The evening before has taken center stage in popular culture. It is no longer widely seen as a quiet day of remembrance of martyrs and holy men and women and all of the dead, but instead as an excuse to get drunk and dress provocatively. I'm no saint, and

I am not wagging my finger at libations or flirtatious outfits, but you can imbibe and dress up any day of the year. All Hallows' Eve was never supposed to be about that. It was so much more, and not just some sober chanting to the dead. October is a party as well— less frat house and more wake; a big ruckus of happy remembrance.

The diehards who fill church pews on All Saints' Day can't take it too personally, either. Halloween didn't steal their spotlight. All Hallows' Eve and All Saints' Day are themselves modern incarnations of an ancient festival that came before them. The pope of the time chose to assign All Saints' Day to November 1, since that date was already a popular pagan holiday. Out in the crofts of the British Isles everyone was already remembering the dead, and they called it Samhain. Samhain, pronounced "Sowen" or "Savan," was a smiling eulogy to family members lost during the harvest year. The folks of Ireland, Scotland, and England were already focused on remembering the dead, and it made a lot more sense for the pope to piggyback Catholic beliefs and holidays on the preexisting harvest and family festival than to tell celebrants that their day was satanic and corrupt. As Christians have been doing for years, the pope figured, if you can't beat 'em, join 'em.

Samhain was one of four big fire festivals on the Celtic calendar and it was considered their New Year. It remained a day of great import and was brought over to America by the Scots-Irish. Eventually it caught on, and over time, All Hallows' Eve became a victim of our modern horror-movie industry looking for a historical scapegoat to justify some juicy R ratings. Tell someone your slasher movie is based on a true story of ancient people's sins or satanic witches and you've got yourself a crowd pleaser. But Samhain was not about Satan, blood, gore, knives, or dead goats.

Samhain was a quiet day, a time to take stock—literally. Most of the farm work was behind them now—people weren't out gathering sheaves of grain anymore, so they could inventory their efforts. Samhain—Gaelic for "summer's end"—marked the end of the entire light half of the year, which started in March. As days grew shorter and temperatures dropped, people started spending more time indoors, around the fire, feasting on the fat of the land and telling stories. Some of the stories were for entertainment and others were for therapy. With the work of growing the winter's food done, there was finally time for a bit of reflection. Folks could slow down and think. And they couldn't help thinking about their own sorrow. If you are alive in October to count sacks of dried beans and your sister isn't because she drowned in June, that hits you with a deeper grief when you have time to reflect than it could when you were so immersed in the labor of the fields.

Thus, a lot of people found time and opportunity to release emotion when the summer ended. All that time spent indoors stewing in nostalgia, thinking about lost loved ones, mistakes, and regrets—that kind of thinking changed a culture and created a holy time of remembrance. It was all around them, this inward reflection. They paid attention to nature, watching her shrivel up and seemingly disappear. Can you picture those early agrarians of northern Europe staring out of cottage doorways at the chill, windy, darkness of early nightfall in autumn? Can you see them closing that door because the harsh cold winds of winter forced them inside with their peat fires? In the highlands, life went from sufficiency to questionable survival. And you can imagine how that quiet, cold time with its focus on remembering the dead had some of the right ingredients to go from solemn to a little bit creepy.

We branded it with creepiness, starting with the early settlers to America, especially in the back-country settlements of immigrants from the Irish and Scottish borderlands who wanted to keep the fires burning on Old Hallowmas to welcome home the memories of their beloveds. They believed with all their hearts that that night before All Saints' Day was the time you could slide into the spirit world—"the veil was thin," as they used to say around the cook stoves and campfires. The doors of communication were open not just for you to talk to your dead relatives in prayers before bed but for them to talk to you.

So naturally, ghost stories and games of fright popped up, and by the 1800s Halloween was headed toward becoming the frightfest we know it as today. "The Legend of Sleepy Hollow," a story I hold dear, was published in 1820, and suddenly the wilds north of New York City became a place of superstition and tradition. Halloween became less and less a pre-Christian night of memories and more and more a night of macabre mirth and parlor tricks.

After the two world wars the first American Halloween parades started in the Midwest and Pennsylvania, and they were a hit. Their purpose was to make Halloween fun again, a time for children, games, and dressing up. The country had time to exhale again and spend money. Halloween was back, as big a deal as it had been in the Colonial era. But nothing alive stays the same for long. Those children raised in innocent community celebrations and neighborhood trick-or-treating grew up to become the filmmakers, authors, and artists who revived the Halloween brand for the silver screen. When the 1970s smash hit *Halloween* turned a cute town celebration into a murder spree, innocence and wholesomeness left the holiday. It seemed as if overnight Halloween plunged from adorable and mythical into moral squalor. It only got worse as the years passed. By the time I was a preteen in the 1990s the horror-movie industry was ready for another revival. Kevin Williamson's *Scream* trilogy was all the rage. Halloween had become a Sodom and Gomorrah freak fest. On October 31 we're more likely to find people dressing up as scantily clad nurses doing Jell-O shots than sitting down to a quiet meal meditating on a photo of a beloved grandmother.

I did not live a thousand years ago. But as a homesteader whose entire paradigm has shifted to one of living in tune with the earth and her seasons, I feel more at home with this ancient version of the holiday than with any modern one. And every year since college I have celebrated in my own way, with my own traditions, meditation, and prayers. I'm not alone, either. Christians, Jews, Buddhists, and Muslims see the harvest season as a period of quiet. All Saints' and All Souls' Day, Rosh Hashanah, Obon in Japan, Eid al-Adha—all of these holy days are about meditation, memories, forgiveness, and starting anew. I like to think that the reason for all this collective introspection has a practical root, that nature's the inspiration. Daylight is, literally, burning away. All the world is folding its wings into slumber; the work that swallowed the summer is behind us. It makes the mind do a funny thing: remember.

My Halloween is mostly about memories. It is nostalgia and gratitude manifest. The work of the past three seasons is behind me—from seedlings to canned harvest—and a world of indoor comforts lies ahead. So when the days get short and my mountain road smells like light gray wood smoke, I can finally slow down and celebrate the way my people have for thousands of years. And when I do celebrate, it is not partying, or dating, or even sleeping that is on my mind: On my mind are the people I love whom I have lost. They deserve the honor of happy memories. The ones we have loved have touched us in ways we can't ever repay or even comprehend, but we can say thank you. We can let them know, wherever they are, that the love lives on.

Samhain has been ours for over six thousand years of our collective human history. That makes it the oldest celebration we still have. Much older than the Sermon on the Mount, or enlightenment under the Bodhi tree. Halloween is the holiday about the two big things: life and death. It's a time for fog-thick memories, silent meals with extra plates set for the departed, and visiting cemeteries to say a prayer. It's a day to be quiet and realize this is all over very soon, and to be okay with that and celebrate what's left with all you've got. Only when you accept your own mortality are you truly free to live this life without hindrance or fear. Halloween is an affirmation of the gift of life. The memories and rituals it brings make me feel lucky, and grateful, and so very, very happy to have the time I have surrounded by people I care so deeply for. I still keep Samhain's warm heart beating in my own. Some things should be left to writhe and sing. They are the more beautiful for it.

NOVEMBER FIRE

ONE CHILLY NOVEMBER MORNING, my friend and riding mentor, Patty, and I were in her new green wagon, sitting on the buckboard and taking turns driving her big gray Percheron, Steele. I had finished my writing work, and when I got a call around noon to come over for a cart ride, I happily accepted. It was a sunny day, and as we made our way fast over farm and field we waved to deer hunters by their trucks going over their game plans and stopped to chat with locals. If you think a thirty-degree day in November is cold enough to keep Washington County residents away from the Battenkill Creamery, well, you don't know us very well. Both of us enjoyed the trip in the wagon. It was a beauty, and as a new driver I was jealous. It had a nice front seat and a bed in the back that could fit two adults comfortably, a load of hay, or any gear you wanted to take to friend or field. Now, on our way home, our bellies were full of ice cream, and we were talking and laughing. It was a happy scene as we neared Patty's home, on Lake Road.

But when we crested Lake Road we saw billowing clouds of smoke behind Patty's historic farmhouse. At first Patty thought that her husband must be home from work and burning trash, but from a half mile away we could tell the smoke was coming from behind the house. Their dog, Harley, was pacing and yelping. This was bad.

Patty had Steele speed up from his Sunday trot into a full-out canter. I was hanging on to the buckboard as she leaned forward to give him a little more chase in his reach. If you have never been on the back of a wagon speeding toward a fire with a ton of horse thundering ahead of you, then you haven't known the true meaning of "make haste." It was wild! It seemed like mere seconds had passed when we were running up Livingston Brook Farm's driveway. As we approached the horse-tack barn Patty just threw the reins into my hands, leapt off the cart, and ran toward the source of the smoke behind the house.

I knew what I had to do: remove the cart from the harness, remove the fifty-pound harness from Steele, and get him away from the house and into the pasture. I didn't know what was happening, so I just went into action mode. When the horse had been freed of the cart and the harness and was in the pasture I ran in the direction Patty had dashed.

To my great relief, she was behind the house, speaking into her cell phone, and carrying a garden hose. The fire was in the woods, not the farmhouse, but it was spreading out in a thirty-yard semicircle downhill. She told the fire department that ashes had been dumped in the woods in the morning; they must have been smoldering and caused dry leaves to catch fire. The wind could blow the foot-tall flames and burning leaves toward the house, which was just twenty feet away. She seemed to be doing all she could, so I decided to go take care of Steele.

The trucks arrived ten minutes later, and a man called Seabass, wearing a yellow uniform and carrying a huge hose, jumped off. He pointed the stream of pressurized water at the flames and had them under control in seconds. It was quite the thing to see. By this time Mark had returned from duck hunting. He was glad to see the house safe. He and Patty chatted with the fire squad, explaining and listening to their assessments, and soon Patty went inside to cut everyone out there a slice of homemade apple pie with a ginger-bread-cookie crust. No one turned it down. Seeing a pack of men in uniform with eagles on their helmets eating slices of apple pie was so thick with Americana I expected to see Harley run around the kitchen with sparklers in his teeth.

Since the fire hadn't been lethal, everyone was in good spirits. Lessons had been learned, and neighbors called to ask about the ruckus. Within an hour of the sudden conclusion of a speeding wagon ride we were all around the farmhouse kitchen table enjoying adult beverages and laughing. It could have been a disaster, but instead it was a story. A story that included ice cream, horses, a farm, and booze, so I was as grateful as a fat tabby on milk truck day.

I couldn't help but wonder what could have happened if Patty and I had been different people with different lives. Our idea of a fun afternoon was a cart ride to a creamery, but what if we had wanted to go shopping in Albany or to dinner and a movie in Saratoga? Had we not been gallivanting so close to home there's a good chance that their 1700s wooden farmhouse could have been reduced to a basement and a chimney instead of kissed by a little bad luck. I have the same protective feelings about my own farm—maintaining proximity, especially in winter when a roaring wood-stove can mean a lot of potential accidents. When you heat your home with actual flames in a metal cage you don't really feel like taking off for a full day, not when you risk a cat or dog knocking something against the stove or, at the very least, coming home to a cold house that requires hours of stoking and care of the fire to get the temperature back up to a reasonable level.

It's a different life up here. But no matter where you live, apple pie and doused fires make for a good night's sleep.

SIGHS

THERE ARE A LOT OF SIGHS on an animal harvest day. You sigh when you hear the first shot of the .22 rifle. You sigh when a life ends moments later, and a body turns still. You sigh as you take stock of what has just happened, and that it was your decision. There is quiet respect paired with a sense of urgency, because you do not have the time to mourn or second-guess. A hog slaughter is a nonstop machine in motion, with steps that must happen in a timely fashion. It's intense, and not enjoyable at all. Yet you do the work and you sigh those sighs and accept them. They are the exhales of those decisions. You own them, and you move on.

In *The Dirty Life: A Memoir of Farming, Food, and Love*, Kristin Kimball writes about how different animals die. How the steers seem to drop with a force stronger than gravity. How chickens flap and seem to panic, and how pigs scream and bleed and thrash. When she started farming, she thought these were the beasts' personalities coming out in their deaths—the calm steer, the quirky chicken, the charismatic pigs—but after a while that assumption died with the livestock as she witnessed more and more deaths. Death was the firing, or ceasing to fire, of a series of synapses and nerves, a chemical reaction that occurs at the end of a life. What I see in the pen on a day of hog slaughter is not a piteous flailing, but a last explosion of life, the mind's finale of fireworks sent through the parts it has always controlled and moved. The struggle was energy leaving the body and moved into another form. A mystery and a gift.

Once still, the animals are brought to our little farming community's abattoir on wheels, the Stratton Truck. Greg Stratton, the butcher, comes highly recommended; good friends who had recently used his services for their family beef steer praised him highly. This was a man well appreciated, and his careful work showed why: he was professional the entire time.

What happens next isn't delicate, but it is honorable. You see the entire process. And when the work is done and the pigs are on their way to the butcher shop, you let out the best sigh of all, happy gratitude and relief that you pulled it off. Slaughtering means a dark day in the life of these pigs, but it is a bright one for the farmer. It represents food, abundance, fuel for local families, and something of value to barter in the coming months.

The butcher's workmen applauded the clean pen and the fact that the only grime and mud on my pair of pigs was on their trotters. I had offered my pigs the largest pen, the most sunlight, and the most well-rounded diet possible. This was a serious validation, and I did not take their praise lightly. I'm always apologizing to guests about the state of my place, which is never ramshackle but never as organized or orderly as it could be. I keep the lawn mowed and the hedges trimmed, kind of. The burdock hasn't overtaken everything, nor has the honeysuckle. I consider these small victories, what with everything else that is going on around the joint, but it would send a suburban landscaper into fits. It's just not slick or refined in any way, and since I make a living writing about farming, sharing stories and photos, and people travel here to learn, I feel they expect a certain level of polish that I never have been able to pull off. So I always expect frowns when folks take a look at the goat-eaten decorative shrubs, the pile of buckets by the front door with the cracked glass, and a house with siding desperately in need of a power washing. They said my place was scrappy, sure, but my pigs and their home were as clean as all get out. I blushed. I was a clean pig raiser, and that counted for a lot more than sagging fences and visible garbage bins.

I did what I could to help the gentlemen, but there wasn't much for me to do besides pile up the heads, skins, and offal I wasn't saving and remove it from the scene. I have lost any squeamishness around this sort of task, not thinking twice about picking up an intestine or lung and setting it aside. Blood is no longer horrific or confusing, but is the living form of the many buckets of water I carried to the pigs. I now know what the smell of a body cavity is like, and it has grown less obnoxious. Today it wasn't bad at all, since not a single piece of offal was pierced or torn. No unpleasant scents of digestion-in-progress wafted around, and since the pigs were off feed for twelve hours previous to the rifle shots, they didn't have any last spoils, either. As far as this line of work goes, it was as pleasant as possible.

The animals' death is not their whole story. It is just how their story ends. I am proud to raise these animals who lived a life of comfort and care. And a few weeks later, when Greg called me to come pick up my hundreds of pounds of pork, I was whistling the entire ride down to his farm. Gibson rode shotgun as we drove the pickup truck, loaded with coolers and cardboard boxes, down to Hoosick. At Greg's shop I helped him carry out box after box of pork chops, hams, bacon, sausage, roasts, loins, leaf lard, and more. It filled the truck, and I drove home weighed down with good fortune. Money is paper, land can never be really owned, but food is true wealth, and with the haul I was carrying I should have been in an armored vehicle.

When I arrived home I made the necessary phone calls and e-mails to the shareholders, the co-owners of the pigs, who split the costs of the animals, feed, and butchering. That night all came to pick up their pig shares—they arrived with their coolers, ice packs, boxes, and lots of festive laughter. For as much work as pigs are to raise, and as serious as the deaths may be, the happiness of handing friends up to fifty pounds of well-raised meat is the most satisfying feeling I have yet to experience as a farmer. I produce yarn, eggs, milk, honey, and vegetables here, but the sacrifice is not equal to that made to raise meat. All those things are given without the loss of sentient life. But a perfectly roasted chicken on a winter Sunday, a rabbit stew simmering in a pot, or a dozen dinner guests fed all the pulled pork they can eat gives a deep sense of primal comfort. Meat is life. It is dense nutrition, powerful energy, and part of the cycle of fertility. It is the blood, bones, skins, and offal that will be composted for next year's garden. My vegetables are serious omnivores, and the life of these pigs is happily continued in a midsummer sunflower or next Samhain's pumpkins. I'll consume those plant's meat and seeds as well, just like the pork or eggs from the animals who fed them. And of course, any food scraps from a pumpkin loaf or an omelet get fed to the next round of piglets. Life never really stops when you live with a seasonal producing farm. It is its own wheel. The pig story starts at one of the spokes and ends at another, but their energy lives on in my body, and the plants, and so on and so forth into all other beings on the farm and in the community. That energy cannot be destroyed, it can only live on. In an occupation as humble as pig farming, I learn some of the secrets of the universe, of the gods' dance, or even of how, when I am gone,

I can only continue in others. I hope I get to be ashes on soil, or compost, or something useful when I am gone and my body is all I have left to give. It would be a shame to stop a hundred thousand stories with embalming fluid, put in a box where the earth cannot touch me. I could never go home. What could be more terrifying?

QUIET LIGHT

I WAS OUTSIDE IN THE WOODS behind the farmhouse with Gibson, searching the hillside for a small tree that would do perfectly for Yuletide. I had a hatchet in my hand and was listening to Tolkien on audiobook. It was the perfect tool and the perfect story for the job. Together, my dog and I hiked just as Bilbo, Samwise, and Bill, the laden pony, adventured out of the Shire. My company was not as grand, but Gibson was doing his best to help me case the joint. It's a thing, isn't it, trying to find the perfect Yuletide tree? My mountainside is mostly sugar maples, with some locust thrown in—a true homesteader's forest. Sugaring and fence posts, that's what those trees signify. Maples here were tapped long before I was born. When I moved in I found piles of sap buckets and spiles in the dump. Clearly those trees had a job to do. I do not yet sugar, but I do have livestock and there is a legacy of old locust posts from farmers who came before me and new ones I have set myself. I'm very grateful for those trees, but both are as naked as baby rabbits come the winter solstice, and I was pining for green.

So I was on a hunt for a tree out in the same forest I hunt deer, rabbit, and grouse in. This tree would be brought inside and set in the front window, covered with white lights, antlers, and wooden crows, and set into a sap bucket full of rocks and water as a makeshift tree stand. I scanned the understory and tried to locate a contender. I didn't want the fat, squat trees you see for sale on the roadside. I was looking for a tree that was struggling. The kind of hard-luck, Charlie Brown type no one ever brings home. This was not because I was trying to be intentionally sordid. I just knew I only had one string of lights, a handful of ornaments, and a small bucket to stand it in. Anything bigger would look barren, but a scrawny tree would look grand.

When I found the winner I chopped it down and cut off the larger branches at the base. With the trunk and hatchet in my left hand and the boughs in my right, I walked home. Gibson ran past me and doubled back to me. Herding trees was a new thing for him and he felt the urge to excel. I shamelessly cheered him on. That dog was such a beacon of light for me. When he stopped dead and whirled around to face me, eyes bright and locked on mine, my heart grew two sizes bigger. His mouth open in a canine smile, snow flurries gathering around his black mane, I wondered how anyone gets through life without a border collie. I glowed inside as warm and bright as the bayberry-beeswax candles in my house. I had bought a whole box at a holiday market from a young couple who make them from wax harvested by local beekeepers. The farmhouse was radiant with the aroma of bayberry. (The bayberry comes from essential oil; it is the rare ingredient not made locally.) You walked in and all you wanted to do was hail wassail and eat figgy pudding.

I'm not Christian but I do celebrate Yule, at the end of December, for many of the same reasons as Christians do. The pine boughs will be hung around the antlers everyone walks under to enter my living space. Red ribbons and ornaments of stags, snowflakes, and sun fill my little farmhouse. Like my mother I overdecorate. Every year she turned my childhood home into a Christmas pageant.

Here Yule, a modern incarnation of older pre-Christian traditions, is just as jolly.

Yule is officially celebrated on the winter solstice, the rebirth of the sun. After all those dark days and nights, with evenings here on the eastern slope of a mountain starting to grow dark by three P.M.—this return of daylight truly is a reason to celebrate. The Wheel of the Year tells the story of the light and dark halves of the year through the battles of the oak and Holly Kings, who fight all year for control of the seasons. Come the winter solstice the Holly King is defeated, and the oak king reigns. The druids used to celebrate this endless battle by means of mock swordfights, or through the symbols of a red-breasted robin taking the place of the dark winter wren. I don't play with swords or catch songbirds, but I like the imagery and the story of the mighty oak finally getting back the light. As soon as the solstice passes, every day grows a little brighter.

LUCEO NON URO
[AN EPILOGUE]

MANY PEOPLE ARE SURPRISED when they find out about my love of Scottish history and language, since I'm not of Scottish descent. This always takes me aback. Does every devout Catholic have to be Roman born? Does every bull rider need to be raised on an Oklahoma ranch? Do you need to be from New Jersey to know how to navigate a shopping mall? Of course not. People are drawn to the lives they want to live, at least the stubborn ones are.

I identify with the old highland clans' history and people. Their stories are full of dramatic fighters and lovers. People who created an agrarian religion that celebrated life without fearing death. People who loved dogs and horses and hunting and music—who told stories, danced and sang, and understood the importance of a hot meal on a cold, rainy day. I love these people. I love their lives, their livestock, and even their miserable weather. I may not be Scottish, but by both deed and elevation I am certainly a highlander.

I joined the Society for Creative Anachronism in order to learn to be an archer. I was told to pick out a name from the country of origin I wished to study, the personality I wished to participate as. I chose Scotland, and I picked the name Corbie Mackenzie. *Corbie* is an old term for "crow" used in traditional Scottish ballads. The name Mackenzie was a nod to the members of Clan Mackenzie, in the novels of S. M. Stirling, who were renowned archers in those books that I came to love.

The Mackenzies have two clan crests: an older one with a giant stag and a war cry in Gaelic as its motto, and another, newer one with a torch-lit mountain and a romantic phrase in Latin as its motto.

Luceo non Uro, "I Shine, Not Burn!"

What a beautiful way to see the world! To choose to be part of light instead of destruction. We live in a culture of victims and anger. We are surrounded by nonstop news and pundits frothing with rage and fear and disaster, crime, and threats. All around us is this fire, this burning. And if you let yourself fall into it you too will be consumed by it. You'll become angry, depressed, unhealthy, scared, and worried. You will stop living the life you were meant to live. Why would you not choose love? Who cares about the fallout?

It doesn't matter if you're a convinced Christian, atheist, pagan, Jew, Muslim, Buddhist, agnostic, or none of the above. If you are reading this with a heartbeat then you are a fellow celebrant because you're alive. Your pulse makes you my brother or sister in a world on fire. We need to help each other shine. We do it through memory, and kindness, second chances, love, and forgiveness. All of us can take a moment to think about what inside us needs to change, and who we lost that we don't want to let down, and to be grateful we're still alive to do those things.

So many people are going to wake up tomorrow and go through their day with the absolute certainty that they will fall asleep that night. Everyone who dies tomorrow will be wrong. There's no guarantee that one of those people isn't going to be me. I hope it isn't, because there are so many more stories I want to tell, and love I want to find. But I don't make the rules. I never expect to die, but

I also never assume I'll live—not since I almost died that sunny day in Tennessee.

Not everyone can make their wildest dreams come true, but hell, everyone *can try,* can't they? Why do so many people choose to put off happiness? Or choose to not try for it? Why do they do things that make them sad? Why do they choose fear and anger and step into the fire that consumes them instead of following the fire that lights the path toward something better?

I can't answer these questions for you. But I know that there's a flock of sheep, a black pony, a loyal sheepdog, and the beating heart of a Mackenzie on my mountain farm. All of it is here because that's what this short, blessed life led me to. I chose to Shine, Not Burn. And that is a choice for all of us. And it shows that life can become whatever you are willing to create.

Now, go light your torches.

ABOUT THE AUTHOR

JENNA WOGINRICH is a homesteader on a small mountainside in Washington County, New York. There she writes, raises livestock, tends a garden, and works with horses, sheepdogs, hawks, and humans. Surrounded by a supportive community and beautiful farmland, Jenna has learned many skills and stories of country living. She's on the south side of thirty and drinks more coffee than she should. You can find her at coldantlerfarm.blogspot.com.